Planning the Night-time City

The night-time economy represents a particular challenge for planners and town centre managers. In the context of liberalised licensing and a growing culture around the '24-hour city', the desire to foster economic growth and to achieve urban regeneration has been set on a collision course with the need to maintain social order.

Roberts and Eldridge draw on extensive case study research, undertaken in the UK and internationally, to explain how changing approaches to evening and night-time activities have been conceptualised in planning practice. The first to synthesise recent debates on law, health, planning and policy, this research considers how these dialogues impact upon the design, management, development and the experience of the night-time city.

This is incisive and highly topical reading for postgraduates, academics and reflective practitioners in Planning, Urban Design and Urban Regeneration.

Marion Roberts is Professor of Urban Design at the University of Westminster.

Adam Eldridge is Research Fellow with the Department of Urban Development and Regeneration, at the University of Westminster.

Planning the
Night-time City

Marion Roberts and Adam Eldridge

Routledge
Taylor & Francis Group

LONDON AND NEW YORK

First published 2009
by Routledge
2 Park Square, Milton Park, Abingdon, Oxon OX14 4RN

Simultaneously published in the USA and Canada
by Routledge
270 Madison Ave, New York, NY 10016, USA

Routledge is an imprint of the Taylor & Francis Group, an informa business

© 2009 Marion Roberts and Adam Eldridge

Typeset in Univers by Wearset Ltd, Boldon, Tyne and Wear
Printed and bound in Great Britain by TJ International Ltd, Padstow,
Cornwall

British Library Cataloguing in Publication Data
A catalogue record for this book is available from the British Library

Library of Congress Cataloging-in-Publication Data
Roberts, Marion.
Planning the night-time city / Marion Roberts and Adam Eldridge.
p. cm.
Includes bibliographical references and index.
1. City planning–Great Britain. 2. Nightlife–Great Britain. 3. Lifestyles–Great Britain.
I. Adam Eldridge. II. Title.
HT169.G7R63 2009
307.1'2160941–dc22   2008052338

ISBN10: 0-415-43617-6 (hbk)
ISBN10: 0-415-43618-4 (pbk)

ISBN13: 978-0-415-43617-5 (hbk)
ISBN13: 978-0-415-43618-2 (pbk)

To Matthew Bennett, central city activist, resident and entrepreneur, with many thanks.

Contents

List of figures   viii
List of tables   ix
Acknowledgements   x

1   Introduction   1

2   Cities at night   10

3   Visions of the night-time city   33

4   Party cities   54

5   Binge-drinking Britain?   82

6   Regulating consumption: mainland Britain   109

7   Regulating licensing: the dream of a continental style of
    drinking   136

8   Planning and managing the night-time city: rhetoric and
    pragmatism   159

9   Consumers   185

10 Night-time cities, night-time futures   207

    Bibliography   224
    Index   241

Figures

3.1 The Ramblas, Barcelona at 10 p.m. This relaxed scene
typifies a more inclusive night-time culture   36
3.2 Digital Heaven   48
4.1 Licensed premises in Leeds city centre   59
4.2 Overlap between different 'economies'   59
4.3 Soho at 8 p.m. The crowds and traffic continue into the
early hours of the morning   68
4.4 Temple Bar, Dublin 2001. Licensed premises   75
5.1 Alcohol consumption in the UK: 1900–2006   97
6.1 Crowds outside a nightclub in London on a Friday night   117
6.2 Terminal hour of 11 p.m. mapped against recorded incidents
of crime and anti-social behaviour   128
7.1 A typical 'lane' in Melbourne, Australia   153
8.1 Organizational chart for Cheltenham Borough Council's
night-time economy strategy   174
8.2 Cartoon, alcohol disorder zone, first published *Observer*,
23 January 2005   178
8.3 Sections through principal streets in four different
micro-entertainment neighbourhoods   182
9.1 Closed clubs appear hostile in the day time   204
10.1 A market in Barcelona at 9 p.m. on a Sunday night   213

Tables

3.1 Corporate ownership of nightlife in the UK, February 2008   51
5.1 Binge drinking by age   89
5.2 Average weekly alcohol consumption (units), 1998–2005   90
5.3 Pure alcohol consumption: litres per capita, 1993–2004,
    ages 15 and over   98
7.1 Selected categories of licensed premises in Barcelona
    and permitted hours   148

Acknowledgements

We have many individuals and institutions to thank for their help and assistance in the projects which formed the background for this book. First, we should thank Matthew Bennett, whose tireless campaigning provided the first and continuing inspiration for our immersion in the subject area. Audrey Lewis gave sound advice on the intricacies of licensing policy over a number of years. Andrew McNeil supported our research into the environmental context of the night-time economy with a wealth of knowledge and an always open mind. Paul Davies and Hannah Mummery from the Civic Trust were our partners and funders for some of our investigations, and their enthusiasm continues. Dr Chris Turner and Dr Galina Gornostaeva were researchers on two of the early investigations and we are indebted to them.

Our studies into the night-time economy were supported by a number of different institutions. We would like to thank the Institute of Alcohol Studies, the Leverhulme Foundation, the Civic Trust, Grosvenor Estates, the Office of the Deputy Prime Minister and the London Borough of Camden for their funding. Our colleagues at the School of Architecture and Built Environment at the University of Westminster kindly gave us some time to write the book and funding for further visits.

It is difficult to know how to acknowledge the large number of public officials, councillors and private persons who gave up their time in interviews and discussions. Special thanks must go to officers from the following city councils in the UK: Westminster, Brighton and Hove, Edinburgh, Norwich, Liverpool, Cambridge, Sheffield, Newcastle and York, and the borough councils of Havering, Doncaster, Greenwich, Basingstoke, Newmarket, Tower Hamlets, Hackney, Maidstone and Chelmsford.

Crown copyright material is reproduced under the terms of the Click-Use Licence.

Outside the UK, we owe officers and councillors from the city councils of Copenhagen, Helsinki, Barcelona and Dublin a debt of thanks. Many police licensing officers provided invaluable insights, and we owe a special thanks to Andy Trotter from the British Transport Police for discussing his experience. Many senior executives, local licensees and managers operating in the night-time economy agreed to be interviewed by us, and some kindly did so twice. The anonymous respondents in our focus groups will never know how they helped us. We would also like to thank David Vijnikka for his help in translation, Mervi Ilmonen and Jonna Kanagasoja for their joint invitation to Helsinki, and Leah Bartczak for all the transcriptions.

Toby Price, Ainsley Crabbe and Robert Thompson (www.robert-thompsoncartoons.com) provided us with a selection of images for which we are exceedingly grateful. Thanks must also go to the World Health Organization for permission to reproduce their table and the British Beer and Pub Association for the use of their graph.

In addition to our formal interviews, we have benefited from our involvement in wider networks that go beyond the boundaries of our own disciplines. Special thanks must go to Professor Guy Osborn and Steve Greenfield from the University of Westminster and Dr Philip Hadfield and Dr Adam Crawford at Leeds University. Andrew McNeill and Phil Doyle provided comments on this manuscript, for which we thank them. Numerous students in our MA seminars have supplied insights and observations. Adam would like to extend a special thanks to Brian Jamieson, and Marion also thanks Piers Corbyn, whose abhorrence of alcohol has provided an interesting opposing view throughout. Finally, we make the usual proviso that the views contained, and the errors, are our own.

Chapter 1

Introduction

we have developed a new economy in this country, a night-time economy....
One of the few areas where we can develop jobs and where we can create
wealth is in the alcohol industry and in the night-time economy – bars, clubs
and industries such as fast food.

Dick Hobbs (House of Commons 2005: Q198)

the current phenomenon of youth drinking and town-centre domination may
well have peaked. This means that alternative scenarios for town-centre
evening life once again become a real possibility.

Ken Worpole (1992: 79)

There is a perception that many town centres across the country have
become 'no go areas': at certain times of the day or week.

(ACPO 2005: 3)

Town centre development hit the headlines in British newspapers in 2005. This
was not because of ambitious schemes by town planners or architects, or
because of town hall scandals, but because a new kind of economy had
emerged. In his evidence to a Parliamentary Committee, Professor Dick Hobbs
identified this new phenomenon as the night-time economy, one which, in its
present form at least, is based on alcohol-related entertainment. Hobbs was
not the first to coin the term night-time economy, but it has taken root and is
now used widely as a catch-all phrase to describe the expansion of nightlife.
Alcohol-related nightlife has expanded in most towns and cities across Britain,
in many cities in mainland Europe, in parts of the USA and in Australia. In Britain
its expansion has been problematic, as the comment made to the same committee from the Association of Chief Police Officers illustrates. The expansion
of the night-time economy has become associated with crime, disorder and

anti-social behaviour, the antithesis of the type of town centre planners and urban designers are trying to create.

Of course, the incidence of drunkenness in town centres is not new. Ackroyd (2001) describes how London has been awash with alcohol at various times in its long history. Each market town delights in the numbers of historic pubs it contains. Although the twentieth century is normally portrayed as one in which the strict licensing laws introduced during the First World War led to sobriety, the 1980s saw a concern with 'lager louts' in provincial towns and champagne excesses by 'hooray Henrys' in the cities (Ramsay 1990). The optimistic quotation from Worpole in the early 1990s came from his analysis of the trends which he and his associates thought would lead towards more inclusive, less alcohol-based night-time activities in towns and cities.

Sadly, their predictions were flawed and the remaining decade-and-a-half saw an unprecedented rise in a youth-dominated, alcohol-soaked night-time economy. That they could forecast another type of future is evidence that the type of night-time economy that has evolved was not inevitable. Moreover, they asserted that the character of night-time activities, including everyday behaviours such as going to the pub, are an important component of the character of a town, and as such are of interest to planners and urban designers. In essence, these are the two themes that have shaped this book.

Perspectives on the night-time economy

A debate on the nature and root causes of the ills of the night-time economy was played out in three issues of the professional journal *Town and Country Planning* in 2004 (Hadfield 2004; Montgomery 2004a, 2004b). The event that stimulated the exchange was the anticipation of the implementation of the Licensing Act 2003 in England and Wales, which, amongst other things, allowed 24-hour drinking and passed the responsibility for licensing from magistrates to local authorities. At that time, it was anticipated that the Act would come into force in January 2005, whereas in fact it was delayed until November of that year. The protagonists in the debate were John Montgomery, a town planner, and Philip Hadfield, a criminologist and licensing expert.

Montgomery defended his involvement in the early expansion of the night-time economy. As a member of a group of urban theorists and cultural consultants, he had argued for the liberalization of licensing laws. This has resulted in what he has since termed the 'tyranny of yob culture', due to both a lack of management of the city at night and to 'something deeply engrained in the psyche of the English which causes them to drink too much' (2004b: 83). Hadfield, in response, pointed out that the cultural theorists who had championed deregulation should have foreseen the consequences of a commercially driven expansion. He went on to argue that blaming the consumer is the 'easy option' that ignores the role the aggressive and powerful drinks and entertainment industries played in shaping demand. In response, Montgomery argued that some local authorities ignored his colleague's recommendations for local strategies to accompany deregulation and that big business could not be blamed for lack of self-responsibility on the part of consumers. He reasserted binge drinking as a peculiarly English problem, in contrast to Germany, France, Italy, Spain, Denmark, Ireland, the USA and Australia. The problems of drunken bad behaviour, Montgomery suggested, lay in English 'mass culture'.

This book explores and extends this debate. The exchanges between Hadfield and Montgomery demonstrate the importance of understanding the growth and evolution of the night-time economy in Britain. The interpretation of its dynamic has a direct impact on the interventions proposed in response. Although there might seem to be an agreement between both protagonists that better management at the local level is required, combined with national guidance, this superficial resolution masks a series of deeper issues, debates and dilemmas. These are bound together in a complex pattern that resists simple, localized formulas.

The unusual status of alcohol as a legalized drug is a thread that draws together many of the issues. As the sub-title of the World Health Organization's report *No Ordinary Commodity* (Babor *et al.* 2003) implies, the problems associated with the spatial and temporal organization of premises that sell alcohol are not the same as those that supply food, or furniture or clothes. The extensive literature produced by health academics provides evidence of the many dangers and hazards associated with excess consumption. Although alcohol is a dangerous and addictive drug, it nevertheless plays a critical role in social rituals in many societies across the globe. The consumption of alcohol is part of everyday life and indeed defines and forms national and local cultures. The 'cultural theorists' represented by Montgomery recognized its importance in underpinning many activities related to town planning, such as eating out in cafés and restaurants, and in entertainment.

Cultural attitudes and expectations have shaped perceptions of 'the night' too. The night is commonly represented and understood according to the opposing narratives of fear and pleasure. Although electrification has diminished distinctions between night and day in developed countries for the best part of a century, the historical legacy of the night's power to transform landscapes and people still shapes activities and behaviours. Flexible working, the expansion of the creative industries and digital communication are gradually blurring previously sharp distinctions between work and leisure and between day-time and night-time. Part of the discussion about the formation of the night-time economy concerns the extent to which activities traditionally associated with the day-time, such as shopping, exercise, education or cultural consumption, for example, visiting art exhibitions or libraries, are able or are enabled to extend into the night.

The extension of the economics of the day-time into spatial and temporal planning for the night provides the sharpest cause for debate with regard to town planning and the night-time city. As the clash between Montgomery and Hadfield demonstrates, the adoption of 'entrepreneurial city' policies by cash-strapped local authorities in the expansion of their night-time economies could be regarded as either misguided, naïve or wilfully negligent. The extent to which the problems associated with unrestrained expansion could or should have been foreseen is not of purely historical interest. If the transformation of previously quiet town centres into 'no-go' areas could have been accurately forecast, then these insights can assist in future planning far beyond the current recession.

However, there is disagreement about the extent to which the expansion of the night-time economy is regarded as a 'problem'. While academic criminologists have no interest in denial, town councils who wish to attract tourists, investment and business into their centres have little appetite for their portrayal as a 'no-go' area after dark. City centre residents are sometimes portrayed as the proverbial NIMBYs (not in my back yard), yet they too have little incentive to reduce the sales potential of their properties. The police are in an ambiguous position too: on the one hand they need to argue for extra resources but, on the other, do not wish to make public their lack of control.

Expansion in nightlife has been further complicated by two partially related trends: the growth of tourism and the extensive corporatization of youth culture and late-night entertainment. Whereas much of the debate in the British media draws on stereotypes of the 'pub' and the British drinker, low-cost travel to mainland Europe has altered expectations and attitudes. An idealized model of a supposedly relaxed 'urbanity' is constructed in debate and serves as a point of reference. Tourism, and the need to enhance its growth, underpins much of contemporary government policy towards consumption. Meanwhile, the shift of youth culture away from illegal raves into 'super-clubs' and other similar venues has re-shaped mainstream provision in the high street. In the drive towards economic expansion, providers have focused on the potential of the youth market, privileging this over more restrained styles of consumption.

In the same way that it is necessary to consider investors, developers and occupiers when analysing the office market, nightlife is shaped by investment, regulation, operators, promoters and consumers (Chatterton and Hollands 2002). The extent to which the problems thrown up by an expansion of nightlife may be ascribed to the structures of ownership and investment is debatable (Chatterton and Hollands 2003; Hollands and Chatterton 2003). The argument that the corporate ownership of nightlife provides a relentless driver for greater profits is compelling. This would suggest, though, that in cities outside Britain, with a different pattern of ownership, the problems may be alleviated or different. Whether problems are caused by a proliferation of venues, or by particular types of establishment, is a moot point. Debate revolves around competition, between bars and clubs packed together in a drinking circuit indulging in price wars with each other, or between supermarkets who use alcohol as a loss leader, thereby encouraging customers to 'pre-load', that is drink alcohol at home, before a night out.

The spatial distribution and configuration of different types of venues has historically been regulated through two different types of regulatory regimes in Britain, town planning and licensing. Changes to the licensing laws have maintained the division between the two sets of regimes, not always beneficially. Both frameworks have to contend with a historical legacy of location and conditions. This would suggest that the problems of crime, disorder and noise can be mainly ascribed to past concentrations, where patterns of provision and consumption were different. Sadly, this is not the case. As the academic studies have noted, many of the problems of excess have occurred in regenerated former industrial zones, as well as historic town centres. Even completely newly built masterplanned districts, such as the Riverside district in Norwich, are not immune. These observations suggest that both planning and licensing need to acquire a deeper understanding of the dynamics of the nighttime economy and the extent to which it can be embedded in different types of urban landscape.

Understanding and representing the views of consumers poses a challenge for researchers. Not only are patterns of consumption varied (Jayne *et al.* 2006), there is also the issue of latent demand (Eldridge and Roberts 2008). Representations of consumers become loaded with moral meaning and connotation, between the 'desirable' image of the restrained theatre-goer and the representation of the 'urban savage'. The extent to which people may be manipulated by an unscrupulous alcohol and entertainment industry is a key point of departure for policy. If the fault does indeed lie with the individual, then anti-social behaviour must be remediated by harsh control, even if this risks criminalizing large numbers of young and not so young people. If the fault lies with ill-conceived or inappropriate deregulation, then the government needs to intervene, even if this is at odds with current conceptions of a self-regulating market and a discourse of partnership between public and private interests. Not only is the 'hidden hand' of the market shaping provision on the ground, there are other equally important variables that shape consumer behaviours, such as class, ethnicity or gender.

In his debate with Hadfield, Montgomery laughs at himself for having proposed as an image of urbanity, 'visions of an elegant café society, with British people strolling about civilized streets as the Italians do – pullover draped over the shoulders, an attractive woman on one's arm'. It goes without saying that this particular image is middle-class, male, white and heterosexual. The point remains about how to construct an alternative vision for nightlife, inclusive of different sexualities, ethnicities, ages and social class. As Skeggs has pointed out, there is a repugnance frequently expressed about 'drunk, fat and vulgar' (Skeggs 2005: 965) working-class women enjoying themselves. Talbot and Bose have demonstrated how the contemporary night-time economy has also privileged white interests, even in areas where there are highly motivated ethnic minority entrepreneurs and audiences (Talbot 2007; Talbot and Bose 2007). The extent to which the night is represented as exotic, a 'liminal' zone in which the notion of normality is reconfigured, is therefore problematic. Such representations are in themselves exclusive, and exclusionary, and deter other interpretations of the evening and night as say, a locus for family excursions or for more mundane activities such as shopping, or going to a meeting of a local interest group.

The manner in which activities at night are produced, bounded, represented, performed and regulated draws in debates from criminology, health studies, sociology, moral philosophy, cultural geography, urban design and town planning. Ultimately, nightlife is experienced through the body, in material places and spaces. This book focuses on the external street environment rather than the dynamics within venues themselves. Hobbs and his colleagues have conducted intensive and high-quality research on door staff, who are the front-line between the two arenas (Hobbs *et al.* 2003). We therefore refer readers to this body of work and instead aim to bring to the fore the issues and aspects of the contemporary night-time economy that will be of interest to built-environment professionals. We explore these issues because we believe that they are important to the future development of 'shaping places' in the construction of everyday life. Increasing urbanization provides a potential for greater use of the hours after dark for activities outside the home. That the ability to be in public at night is so constrained is a matter of regret. The limitations and self-limitations that result in only 15 per cent of over-55s venturing into town centres (ODPM 2005c) after dark is surely a loss to society, to human experience and to urbanity in its wider sense.

Furthermore, we are sympathetic to the aims of the government's policy for an 'urban renaissance' that seeks to encourage town and city centre living. The expulsion of housing from many British town centres can be attributed to the norms and practices of modernist town planning, with its designation of mono-functional zones that dispersed a more vital mix of uses into soulless housing estates and industrial parks. In turning the clock back to nineteenth-century patterns, old conflicts have re-emerged. Whereas in the nineteenth century complaints were heard about the noxious fumes and noise emanating from factories, now in the twenty-first century residents complain about the noise and disorder coming from the streets as the new 'drinking factories' disgorge their clientele at 3 a.m. into their waiting coaches and cabs.

Of course, it might be argued that it is not only the production of the built environment that shapes places at night. Other factors play an important part, as has been recognized by the town centre management movement. Transport, parking, street cleansing, lighting and policing all contribute to the experience, as does the detailed design of the public realm. The significance is not purely local, for many major nightclubs have a regional catchment. As in the day-time, issues of payment for extra services arise, for extra policing and security, cleansing and so forth. Whether these should fall on the shoulders of the public or private sectors leads back to the discussion of 'boosterism', of the morality of encouraging the over-consumption of alcohol, of the way in which 'the night' is shaped and framed by policy and regulation, and of the rights and responsibilities of the individual.

In concentrating on the mainstream, we have had to ignore other activities traditionally associated with 'the night', such as sex work and illegal drug use. This is partly for reasons of space, but mainly as a result of the empirical research that has provided the justification for writing this book. This research has focused on environmental management and has largely been of a practical nature. It is to this that we shall turn to next.

Empirical research

The research projects on which this book is based have been localized, and have investigated the themes raised by licensing deregulation in England and Wales. The investigations have drawn upon Chatterton and Hollands' (2002) framing of the night-time economy in terms of producers, regulators and consumers.

A study of Old Compton Street in London's Soho used time-lapsed video- and audio-recording techniques to identify the extent and nature of the disruption caused by the numbers of people on the street throughout a summer weekend in 2002 (Roberts and Turner 2005). The results were analysed through pedestrian counts and represented in graphs and charts. A similar technique for analysis was used in an unpublished study for an inner-London borough which had commissioned its own observational studies of pedestrian flows and incidents of anti-social behaviour. We also used direct observational techniques as part of our study of five night-time economy 'hot spots' in different town centres in England. Here, rather than video recording we conducted an observational study in the micro-district once every hour from 8 p.m. until after all the clubs had closed and the streets were deserted, taking notes of people, activities and incidents.

Consumer attitudes were probed through focus groups. One study investigated five towns in England with four focus groups in each, a total of 20 groups that included 160 respondents. The composition of three of the groups was common to each location; young workers, the carers of young children and more affluent residents. The fourth group in each location responded to its particular circumstances, and in one place was composed of respondents drawn from black and ethnic minorities, and in another from younger women. Another study for a market town in the south-east of England drew on material from six focus groups, comprising residents of different income levels, parents and students. A further focus group was drawn from an affluent commuter village adjacent to the town. Our consumer research differed from the studies conducted by researchers interested in youth culture, criminology and health, as our object was to investigate what deterred respondents from going out at night. It therefore included a wider range of people and, building on work carried out by researchers in Cardiff and Swansea (Bromley *et al.* 2001; 2003), probed attitudes and experiences.

The attitudes and experiences of English and Welsh local authorities prior to the introduction of the Licensing Act 2003 was investigated by a postal survey in 2003–2004 (Roberts and Gornostaeva 2007). This provided a snapshot of the extent to which local authorities were reporting an expansion of their night-time economy. The survey justifies the focus of this book, in that the majority of local authorities who reported problems associated with expansion cited issues such as lack of transport and general feelings of threat as of greater significance to them than fighting and violent assault. A wider view of the regulatory environment was taken in three further studies. The first investigated British arrangements in comparison to three other northern beer-drinking countries, and investigated the West End in London, Temple Bar in Dublin, the Hackescher Höfe in Berlin and Nyhavn in Copenhagen. This study drew on interviews with licensing officials, the police, planners, environmental protection officers, fire officers and residents' representatives (Roberts *et al.* 2006). A similar pattern of cross-agency interviews was used in a later study of four urban locations in the eastern region of England as a follow-up study after the implementation of the Licensing Act 2003. In this we added transport providers to the list.

We have conducted two surveys of the corporate producers of nightlife. These, as will be discussed, operate across a national sphere of operation. There are also other significant stakeholders in the night-time city, businesses who operate at night, such as supermarkets and convenience stores, but who would resist a definition that incorporates them into the night-time economy. Different types of entertainment providers similarly resist such a categorization, such as the owners of cinema chains, bingo halls and theatres. We interviewed 23 of these stakeholders, representing 17,544 businesses, in 2004 and went back to 15 of them in 2007 for further semi-structured interviews. Local practices and experiences differ from those offered by head office executives, and our follow-up interviews with local managers in five 'hot spots' captured the views of some 40 of them.

This primary research provided a basis for the production of this book. Three other non-commissioned surveys also helped to widen its scope. Telephone interviews with five hen and stag party providers gave an insight into the operation of that new sub-section of the hospitality industry. Visits to Edinburgh, Barcelona and Helsinki that included interviews with local 'stakeholders' in each location broadened our geographical scope. Telephone interviews with planning officials in Copenhagen updated our information, as did correspondence with regulatory bodies in New South Wales, Australia.

Structure of the book

The book starts, in Chapter 2 with a discussion of the night as a timezone that is framed by different perceptions and narratives. It contrasts these abstract and extreme tropes with the extension of everyday pursuits into the hours after dark, thereby providing an introduction for the central narrative of the book, the expansion of the night-time economy. Chapters 3 and 4 chart the context and dynamic of this expansion, including the influence of tourism in the rise of 'party cities'. Chapter 5 provides further context by considering the health impacts and definitions of alcohol consumption. It tackles the argument that 'binge' drinking is a particularly British problem. Access to the purchase of alcohol, either through consumption in pubs, clubs and bars or in supermarkets is a topic that is now widely debated, as is pricing.

Chapters 6 and 7 move on to licensing. Chapter 6 provides an account of the changes to the licensing system in England, Wales and Scotland. Often heralded as an experiment in deregulation because of, in England and Wales, its allowance for 24-hour drinking, it is actually far more complex. Crude comparisons are often made to the arrangements in other countries, especially mainland Europe. Chapter 7 sheds light on various misapprehensions and contrasts two cities with highly regarded public cultures, Copenhagen and Barcelona. The chapter goes on to consider arrangements further afield, in Australia and across Europe.

Chapters 8 and 9 return to Britain, first considering interventions in the management of the night-time economy in the context of competing discourses and rationales. Chapter 9 draws on our research on consumers and draws some conclusions for policy. Finally, in Chapter 10 we review our arguments and make suggestions for future policy, practice and research.

Chapter 2

Cities at night

In the symbols and dreams of most cultures, night is chaos, the real of dreams, teeming with ghosts and demons.

(Schivelbusch 1988: 81)

The labourer needs time for satisfying his intellectual and social wants, the extent and number of which are conditioned by the general state of social advancement. The variation of the working-day fluctuates, therefore, within physical and social bounds. Both these limiting conditions are of a very elastic nature and allow the greatest latitude.

(Marx 1977: 225)

Introduction

This chapter paints a broad picture of the night and examines the 'conditions of possibility' for the night-time economies of the late twentieth and early twenty-first centuries. Far from being an organic process, the discussion that follows argues that specific forces have shaped the late-night cities and town centres we now have; processes related to technology, the introduction of new forms of leisure and changing patterns of activity and household structure. These and other factors have shaped the development of city centres and contributed to their current perception as vast 24-hour 'entertainment machines' (Clarke 2004).

Rather than being a passive or natural phase that occurs between dusk and dawn (Williams 2008: 514), the night, as it is discussed here, is rendered as owing to a complex and intertwined array of historical, local and global, economic and social trajectories. As will be examined, the ways in which the city after dark has been both consumed and produced is by no means universal, and its history has not followed a linear or singular path. Our experience of the city at night is dependent upon a range of factors; age, class or gender are primary, but so is where and how we live, our economic, social and cultural capital (Bourdieu 1992), caring responsibilities or access to transport (Lucas 2004). Historically, the meanings attached to the late-night city have therefore also changed over time, for different individuals.

In recent years it has become common to associate late-night culture with alcohol and drinking. Alcohol has most certainly played a central role in the night-time city (Barr 1998; Schlör 1998), as this book examines. However, as Jayne reminds us, the relationship between cities and consumer practices are mediated via an array of macro and micro processes (Jayne 2006: 18). Rather than having a direct causal link, the late-night city has emerged in the context of local and international forces pertaining to the economic, cultural and political. New consumer groups and subcultures have played an equally significant role, with these factors further 'constrained and enabled' (Jayne 2006: 19) in relation to the everyday performance of race, age, class, sexuality and gender.

Echoing this point, Latham argues that cities are more than just sites of consumption. They serve as backdrops through which different ways of inhabiting and 'doing' the city are enabled (Latham 2003: 1710). The economic and commercial functions of the city are fundamental to the development of the contemporary late-night city, but the ways it is inhabited, what it means and what people 'do' is far more than just a simple economic relation. As Bell has similarly argued, to focus excessively on economic factors results in ignoring the more 'mundane' activities of sociality (Bell 2007: 3). In short, the city at night is every bit as complex – and can be every bit as mundane – as the city in the day-time (Amin and Thrift 2002).

As such, there is no singular night-time or evening economy, but rather a number of different economies running side-by-side, in support or opposition to each other. This is as true in terms of the ways different cities function at night, as it is for different parts of the very same city. Across the major late-night capitals such as London, Paris, Berlin or New York, there are micro-districts where citizens pursue different pleasures and desires. Streets busy with chain bars and young people may well be adjacent or nearby to streets that play host to families dining in a more relaxed environment. Further afield one may find underground bars, factories teeming with workers, back streets empty and seemingly derelict, or parks that play host to street drinkers or the under-age. A central point of this chapter is, therefore, that the late-night city is a diverse and complex 'time-space'. It is a site where competing interests and desires can rub up against one another, or seemingly exist side-by-side, with each party oblivious to the other. Equally, how we experience the nocturnal city owes to trends resulting from local, national and global trajectories.

To focus on the 'everyday' workings of the nocturnal city may signal a domestication of late-night culture. But, far from the transgressive potential of the late-night city or the pleasure/fear dialectic through which it is commonly understood, the night-time city of today enables new, everyday uses of space. Many of these changes have occurred in relation to broader intense shifts around gender, domestic life and the organization of work. In noting the variety of activities that take place at night, the cliché of the romantic late-night city populated by scoundrels, lovers and pleasure seekers becomes increasingly that; a cliché. The contemporary late-night city is one where the everyday and the spectacular now exist alongside each other. The mainstream media, in recent years, has come to focus almost exclusively on the latter; but less extreme activities at night are no less worthy of investigation than underground rave parties or excessive drinking.

Bearing in mind that the city at night functions and is experienced in these multifunctional terms, it is important to avoid creating a simple opposition here between the mundane and everyday and the more romantic and adventurous. These two aspects of the nocturnal city typically function side-by-side, with late-night workers supporting and servicing the more adventurous activities. Moreover, Phil Hubbard, following Benjamin and Baudelaire, has remarked that it is important not to lose sight of the ways that the everyday is in itself a paradox. It can be repetitious, but it can also be the site for resistance and change (Hubbard 2006: 100). A late-night bartender waiting for a bus on a busy Friday night, a social group meeting in their local pub or work colleagues having their usual after-work drinks – each can entail uncertainty, creativity, spontaneity, chance encounters, contradictions and tensions (Stevens 2007: 18–19). As Lovatt and O'Connor remark:

> the possibility of connecting pleasure and civic space, the attitude to cities, crowds, the mingling of classes, the anxiety of shifting identities, of the charlatan, the cheat, the social climber, 'the dukes for a night' (Clarke, 1985) – all these represent histories as yet unwritten.
>
> (1995: 131)

Perception and defining narratives

If the city at night is forever in flux, a contested and dynamic time-space, the ways it is typically represented owes a great deal to a shared history and perception. The most common trope through which cities at night are understood – as places of either danger or pleasure – remains steadfastly intact. The nocturnal city, as Amin and Thrift (2002) note, continues to be represented in terms of the familiar tensions of pleasure and fear, regulation and chaos, disorder and control. These oppositions reverberate today, harking back to centuries past, when darkness represented danger and threat, but also the promise of illicit adventures. In his discussion of Britain's evening economy, Bianchini (1995) refers to Thomas Burke's *English Night-life: From Norman Curfew to Present Blackout.* Though published in 1942, Burke's characterization of the tensions that encircle the late-night city continues to hold some truth today. Burke suggests that Englishmen (*sic*) have divided attitudes towards 'the Night', with some experiencing a 'shiver of delight' having never got over the excitement of 'Sitting Up Late'. Others, including 'most officials', take the view that entertainment that happens at night is 'verging on the unholy'. The role of authority has been to impede going out with 'authority ... making things as difficult as possible for them, by devising budgets and laws and bye-laws' (cited in Bianchini 1995: 123).

This 'schizophrenic' attitude towards nightlife, as Bianchini terms it, continues to structure our thinking and experience of the city at night. Pleasure and chaos, fear and excitement remain dominant stories. This is, of course, a romantic and clichéd vision of the city. Moreover, it is always important to ask for whom does the city at night represent these twin narratives? As Schlör argues, the story of the late-night city has traditionally been told by police and missionaries (Schlör 1998: 14). In contemporary times, stories about the city at night are told by hospitals dealing with the drunk and unwell, journalists and those free to partake in the city's pleasures. Rarely do we hear the story of the late-night city recounted by licensing officers, the lower middle-class, families or parents. These voices, in one respect, may imply a kind of domestication of the city at night, an image which is anathema to the mystery and anonymity it traditionally represents. Here, the discussion turns to this dominant tension – pleasure and fear – in order to explore how and why the city at night continues to be framed as such, and to draw out some of the key moments in the history of the late-night city.

The expansion of lighting

The implementation of street lighting in seventeenth-century France is, for Schivelbusch (1988), a defining moment in the history of a specifically modern late-night city. New means of lighting not only enabled the expansion of day-time activities into the night; with the advent of gas heating late-night venues could be economically warmed as well (Parkes and Thrift 1980: 328), an especially important requirement in the cooler months. Until mass-produced and economical forms of lighting emerged, light in the form of candles and oil lamps were associated primarily with places of work or the home. The Industrial Revolution required new forms of lighting, however, and, as with all great inventions, the history of providing light for the legions of late or early-morning workers is replete with success and failures (Schivelbusch 1988: 9). In 1770, Antoine-Laurent de Lavoisier, the so-called 'father' of modern chemistry, discovered that lighting relied as much on carbon as air. This led the way for the beginnings of larger lamps, and in turn, modern-day lighting technologies. When, in 1783, the chemist Francois Ami Argand extended de Lavoisier's findings by developing a hollow and movable wick, this allowed for the wick to be raised or lowered resulting in a stronger flame, producing a brighter and cleaner source of light. As noted by Schivelbusch, this groundbreaking invention 'was to the nineteenth-century household what the electric light bulb is to that of the twentieth century' (Schivelbusch 1988: 14).

Despite these breakthroughs, lighting of this period was used primarily in the industrial sector, suggesting it was the needs of industry and commerce, rather than leisure, that prompted the introduction of better forms of lighting. Indeed, as noted by Melbin, Karl Marx in 1867 observed that late-night employment represented an entirely new form of worker exploitation (Melbin 1978: 3), a point we return to later. On the street, the situation was also rapidly changing. Until street lighting was developed, those who ventured out at night did so armed with personal lamps or with the assistance of employed escorts. These 'link boys', in London, or 'falot', in Paris (Melbin 1978: 11), eventually made way for stationary lighting in the form of street lamps, which, though of little comparable value to the lamps later developed by Lavoisier or Argand, began to emerge in Paris in the sixteenth century. From the eighteenth century onwards, the growth of street lighting was rapid. Schlör notes that from an initial 203 lights in Paris in 1835, within four years the number had grown to 12,816 lights on 6,273 lamp-standards (Schlör 1998: 59). Ackroyd explores the comparable implementation of lighting in London. Beyond security measures, lighting, like elsewhere, played a crucial role in the city's commercial sector. The journal writer James Boswell, in 1762, took note of 'the glare of shops and signs' in the British capital (Ackroyd 2001: 442).

By 1807 gas had replaced oil on the streets, with Whitecross and Beech Street, near to London's contemporary Barbican Centre, being the first to implement the new technology. Other streets soon followed, with Pall Mall doing so the following year and Westminster Bridge by 1812. Domestic use followed by 1840 (Ackroyd 2001: 442–4). At the time it was hoped 'the new gas-lighting would not only banish vice and crime from the streets, it would also materially increase the speed and volume of trade' (Ackroyd 2001: 444). By 1878, electric lighting was introduced, initially on the Thames Embankment and, within ten years, to the commercial signs of Piccadilly Circus.

For Lovatt and O'Connor, the coming of light represented more than just the illumination of industry and the means for a growing late-night leisure and commercial sector. It represented the 'control' of nature itself. As a central narrative to modernity, the control of nature, whether that be by the arcades that inverted the indoors and out, the shifting of rivers to build canals or 'annihilating time and space' by means of the new railways, the modern city was illuminated almost as a direct challenge to the superstitious fears of the archaic and medieval world (Lovatt and O'Connor 1995: 131). The night, in other words was to be conquered in much the same way as other natural forces. Lighting was therefore instrumental in the history of modernity, industry and commerce. It was central to the very notion of a modern city, providing security on the city's streets, displaying wares from across the globe in brightly lit windows and enabling the modern resident to partake in later endeavours after dark. If light brought people on to the streets and enabled them to work or seek pleasure in new ways, however, the night continued to be a time-space that spoke of darker fears and desires.

Late-night fears

> Both land frontier and the nighttime have reputations of danger and outlawry. Interestingly, both do not live up to the myths about them, for the patterns of aggression are selective and localized.
>
> (Melbin 1978: 10)

Williams, in his discussion of late-night spaces, draws upon Deleuze's notion of deterritorialization to explain the manner in which space, at night, shifts in meaning, form, use and purpose (Williams 2008). By virtue of darkness, space at night can be destabilizing and comes with a whole different set of social relations, exclusions and inclusions to the day. Different technologies and mechanisms of surveillance operate, and different parts of the city become less or more active, with demographics that may not be as present during the day. While we have suggested that the everyday is increasingly moving into the night, the city after dark continues to be 'approached and appropriated' in contrasting ways to the city in the day-time (Williams 2008: 517). Moreover, as much as the late-night city represents, and continues to represent, a hidden world of pleasure, transgression and adventure (Palmer 2000), it is equally a site of potential danger and threat – two extremes that may not necessarily be opposing.

In medieval times, the coming of night set into force a battery of common practices and regulations, from locking the city gates to locking and bolting one's own home. In Paris in 1380, for example, a decree stipulated that all houses were to be locked and all keys given to the magistrate. No one was allowed to enter or leave a house without a valid excuse. Those who did venture out were often required to have a lamp with them, in case they should be suspected of misadventure (Schivelbusch 1988: 81–2).

Following Williams' notion of 'channelling', which refers to the direction of activities into 'appropriate places', the advent of lighting zoned the city into spaces for site-specific activities (Williams 2008: 522), with other attendant regulations and mechanisms of control and surveillance. As street lights allowed greater numbers to partake in the nocturnal city in relative comfort, so too complaints increased about areas that were not sufficiently illuminated (Schlör 1998: 65). By 1829, the Metropolitan Police was formed in London, largely to manage the bourgeoning late-night city (Reynolds 1998). As a modern-day night-watch, 'the Met' were newly responsible for ensuring the safety and security of late-night wanderers, pleasure seekers and workers. This is not to imply that they were overly successful. As today, policing is central to debates about criminality and vice, and as Schlör argues, London in the 1880s, along with Paris and Berlin, possessed an acute sense of the night as a place of criminality and immorality. This lies in contrast to the London of a century earlier which, as Cruikshank has argued, was noted for its sense of order (Cruikshank and Burton 1990).

In examining the ebb and flow of fears and dangers that surround the city at night, it is important not to see contemporary problems, such as drunkenness and anti-social behaviour, as a recent occurrence. The current discourse about 'binge Britain' appears to imply that until the 1990s, British cities were marked by an all-inclusive bourgeois sobriety. A recent media campaign in Australia has seen a similar association occur, with news headlines in Melbourne (Mitchell 2008; Rood 2007) drawing attention to the myriad problems residents face when venturing out after dark. While not wishing to make light of the violence and anti-social behaviour taking place in Melbourne, or any city centres, the night has long been associated with potential danger, transgression and violence.

Schlör, commenting on how darkness can appear threatening because of its unstructured nature, asks if the 'night was essentially about *fear*?' (1998: 58). Fear is always historically and culturally specific, and not a universally shared emotion. As much as there are different fears, so too there can be different intensities of fear associated with different parts of the city. The city at night may be avoided, not because of the drunken crowd, but because of the barrenness and lack of sociality. That is, whereas the mythical city at night is replete with bawdy adventures and illicit activity, there is an alternative landscape today that is empty and windswept. In Britain, Australia and the USA, central business districts are especially likely to be abandoned after dark, and these pose a whole new array of anxieties (Oc and Tiesdell 1997).

Melbin argues that, traditionally, the nocturnal city is a place of strangers, each eyeing the other for the likelihood of potential danger. Once one decides 'the other' is a potential friend, mistrust turns to camaraderie

(Melbin 1978: 13). He characterizes this tendency towards sociality as due to the fact the city at night is less harried, with 'less places to rush to'. Preoccupied by work or day-to-day appointments, the city in the day has long been perceived as a place of anonymity and aloofness. At night, by contrast, a shared sense of commonality may override our fears. In these terms, it comes close to the characterization of cities as 'enormous machines for the generation of connections between the unexpected and the unexceptional' (Latham 2003: 1719). Williams takes a comparable approach, arguing that the city at night allows for new configurations of connection between groups that do not exist during the day. Neighbourhood watch, different alliances between corporations and local government, disparate women coming together in the form of 'Reclaim the Night' marches; each allows forms of cooperation and coexistence that do not function during the day (Williams 2008: 521). Nevertheless, the city at night has never been entirely accessible to all, 'free of confusions, conflicts, violence' (Harvey 2004: 236). As it does today, late-night culture historically came with a unique set of restrictions, boundaries, regulations, exclusions and borders dependent upon location, racial, class and sexual politics.

Beyond variations and different intensities of fear, as Burke commented, the nocturnal city is represented as a place of 'moral' danger. Lovatt and O'Connor argue that as a site of liminality and transgressive behaviours, the night has always entailed hidden dangers; mythical or otherwise (Lovatt and O'Connor 1995). Moreover, though referring to the 1960s, they point to how the fear and sense of danger that circulates around the nocturnal city is not specifically to do with the night per se, but rather a distrust for urbanity and the people who frequent the city after dark. They argue that 'the liminality of nightlife turned into the pathologisation of city centres, riven by those residual groups who used the city – youth, prostitutes, drug addicts etc' (Lovatt and O'Connor 1995: 132).

Houlbrook's *Queer London* (2006) examines this argument in greater detail, painting an evocative world of the lesbians and gay men who inhabited London in the late-Victorian and Edwardian period. In similar terms, for Schlör, the night is always associated with morality. Drugs, alcohol, gambling, same-sex desires and prostitution have long been central to not only that which repels some from the city, but that which may attract others in equal measure. That the night offers the potential for transgressive behaviours is a rich vein within discussions of nightlife. Measham's (2004) work on hedonistic cultures, or Wilson's (2006) thorough investigation of youth cultures and the Toronto rave scene represent recent examples of how the night is articulated with underground or subcultural uses.

Late-night pleasures

If lighting supported the development and extension of the modern night-time economy, at its heart was a shift from purely commercial or industrial needs to the beginnings of new forms of leisure and play. As Stevens notes, leisure and play differ markedly, in that leisure presumes an organized time and space distinctive from the times and spaces associated with work. Play, on the other hand, relates more to the spontaneous, disorganized affects of urban life (Stevens 2008). If leisure is about the organization of time and space into separate and appropriate cycles, play is what may eventuate from the less structured pursuit of desire. Nineteenth-century forms of the newly organized world of leisure could entail trips to pleasure gardens and the theatre, or far less 'moral' pursuits such as visiting opium dens, brothels or bawdy drinking venues. Leisure, as today, also entailed more playful outcomes, such as chance encounters in darkened alleys.

Schivelbusch argues that 'the baroque culture of the night spawned modern night life, which, since its conception in the cities of eighteenth-century Europe, had grown into a characteristic feature of present-day urban life' (Schivelbusch 1988: 138). Festivals, which remain today a popular leisure pursuit, in the Renaissance and Middle Ages typically occurred during the day. By the seventeenth century, festivals increasingly occurred after dark. Burke's *English Night Life* (1943) provides a rich and compelling history of other festivities and leisure forms from the Norman period until the Second World War. He recounts how theatre performances were, until the seventeenth century, performed in the afternoon. By the eighteenth century a more familiar image of the nocturnal city begins to slowly emerge, one centred around inns, a limited number of theatres, private parties and clubs.

Somers, in relation to the expansion of leisure in US cities in the nineteenth century, follows a similar trajectory and argues that leisure spaces offered one of the few spaces where people of different class and status would truly come together (Somers 1971). This theme has featured strongly within urban studies, and Haussmann's restructuring of Paris and its attendant creation of *grands boulevards*, cafés and bars has provided a rich source of comment with regard to the intermingling between classes that the 'opening up' of the city offered (Clark 1999; Harvey 1985).

In similar terms to Somers, Barr suggests that industrialization and urbanization resulted in 'weakening traditional sanctions on conduct' (Barr 1998: 138). By this he means the anonymity of the city, and how being away from traditional networks of families and friends in the urban metropolis, led to a weakening of the binds that censure behaviour. Raban's (1998) *Soft City* provides an enduring account of the potential that the city offers for anonymity and a reinvention of self. This theme is taken up by Melbin, for whom the pleasure afforded by the city at night is related to various factors, from freedom from harassment and the crowds of the city during the day, to feelings of ownership for those who may traditionally feel excluded from 'normal life' (Melbin 1978: 8). Melbin's claim that the 'ugly, obese and homosexual' may typically be the type to find comfort in darkness is dated, but he is perhaps correct when he notes that the nocturnal city 'serves an insulating function that averts possible tensions from unwanted encounters' (Melbin 1978: 9). Though this obscures the problems that women, and indeed homosexuals and others may encounter at night, the night is a time-space where wanderers and explorers can be free from the rules of propriety or surveillance that operate in the day-time.

A comparable observation has been made recently by Valentine *et al.* in relation to urban/rural drinking practices today. Young people drinking in local bars frequented by members of their local community self-monitor their behaviour in ways they do not when in large anonymous bars in the city (Valentine *et al.* 2008). Kneale (1999), in comparable terms, examines the history of the Victorian pub, principally in terms of how new technologies of surveillance served a primarily moral purpose. The open or long bar, as found in the majority of licensed premises today, was one such example.

Returning to Somers' discussion of new forms of sociality, he refers to the period of 1820 to 1920 as a 'revolution', rendering this period in terms of a receding Protestantism, which had always valued hard work over pleasure. In the USA, as in mainland Europe, the shift from a rural and agricultural society to an urban and industrial one impacted in significant ways on the organization of leisure, and corresponding pursuit of pleasure.

In Britain, one such venue for the association of different groups of people were coffee houses, gin palaces and pubs. In London, the number of bars and pubs increased exponentially in the eighteenth-to-nineteenth century. Ackroyd recounts how Thomas Brown noted of London, in 1730, that 'to see the Number of Taverns Alehouses, etc. he would imagine Bacchus the only God that is worsipp'd there' (Ackroyd 2001: 461). Of course, London has a long history of alcohol consumption. In the thirteenth century the city was renowned for drunkenness, as it is today. A century later, 354 taverns and 1,333 bars were operating (Ackroyd 2001: 346). In the 1740s to 1750s there were an estimated 17,000 gin houses alone, this being the height of London's gin pandemic. Across the UK in 1830 there were 51,000 licensed venues, with a further 46,000 by 1838 (Kneale 1999: 334). Though not specifically serving only at night, by 1870 a total of 20,000 public houses and beer shops were recorded in London, catering to a total of 500,000 people per day (Ackroyd 2001: 357). Pubs were more than just dens of drunkenness, of course, and it should not be presumed that the entire nation visited pubs only to become intoxicated. Moreover, as Smith details, the number of licensed premises per person dropped significantly between the Victorian and early twentieth-century period. From a high of 168 persons per on-license in 1831, the number grew to 595 persons per on-license by 1951 (Smith 1983: 369). The size and capacity of the actual venues may account for some of this, but, as Kneale has remarked, the nineteenth-century pub was more than just a site for consuming alcohol, but instead served a broad social function. Kneale refers to the pubs of the nineteenth century as 'refuges'; places to play cards or dominoes, to read, eat, use the toilet, or avoid one's spouse or family (Kneale 1999: 334).

Despite the variety of activities that occurred inside licensed venues, it has been widely commented upon that since at least the 1600s, pubs have been the subject of moral crusades. Leisure institutions became a cause for general concern in the early nineteenth century, but the pub had been associated with disorder for centuries. The same complaints recur from 1600 onwards, focusing on disorder, sexuality, pauperism and threats to family life (Kneale 1999: 334). By the nineteenth century, pubs were already licensed, but the driving forces behind increased legislation and control shifted 'from economic to social and political motives' (Kneale 1999: 334). Licensing hours, in particular, were introduced, largely as a result of industrialization and the necessity for punctual and sober workers (Barr 1998: 137), a theme we return to in Chapter 5. While this clearly had an economic rationale – inebriated workers not being conducive to productivity – the regulation of when people could and could not drink was intimately entwined with a growing sense of appropriate and inappropriate behaviours, family life and a distinction between work and leisure, both spatially and temporally.

As was the case in Western Europe, as well as Australia, shorter working hours, urbanization and a higher standard of living all contributed to the rise of leisure opportunities in the USA's burgeoning cities. As elsewhere, those opportunities were varied, and ranged from traditional late-night activities such as gambling and drinking, to newer forms of leisure such as the live arts. The theatre, as one example, was traditionally the domain of the aristocracy, but post-1820 it was increasingly targeted towards the middle- and working-classes (Somers 1971: 127). This 'democratization' of the theatre did not equate with the imposition of high cultural values, however. Eating, drinking, chewing tobacco and procuring the services of prostitutes would all take place concurrent to the action on stage, and it was not until the later emergence of the minstrel, variety and burlesque shows that high and low culture, and audiences, began to separate (Somers 1971: 130–1). This defining feature of Victorian nightlife finds echoes today in the association between pleasure and immorality.

Beyond pleasure and fear

Pleasure and fear remain common tropes in conceiving the city, both historically and today. It has been argued thus far that both models have their limitations, however. Pleasure and fear may help to organize the story of the city at
night, but there are alternatives. In the second section of this chapter, it is the
'everydayness' of the city at night that is explored. The city, in the discussion
that follows, is less about pleasure and fear, than about work, new forms of
community and activity at night, the 24-hour society and the rise of single
households. These factors are considered as key aspects of the continuing
history of the nocturnal city.

Two important names in the study of the 24-hour city are Murray
Melbin, as previously mentioned, and, more recently, Leon Kreitzman (1999).
Melbin explains the growth of late-night culture in comparable terms to the
expansion into the wild-west, as if the night were a vast unchartered territory
just waiting to be uncovered and colonized. The 24-hour society, Melbin suggests, emerges in three distinct waves. Phase 1 involves the wanderers, those
who shirk the moralizing discourses of the day and venture out into the quiet
and empty streets to be alone or find similar spirits. The explorers are followed
by Phase 2, late-night services and workers: the waiters, charity workers and
ambulance drivers. Finally there is Phase 3, the consumer sector, which fuels
the emergence of large-scale restaurants, bars and the like (Melbin 1978: 7).
This final stage is where the community of late-night wanderers expands to
include a wider array of people, all demanding a higher level of services and
infrastructure. Unlike the first explorers, content to walk in the darkness and
drink their poor-quality coffee in disreputable venues, the final stage sees a
demand for venues and experiences that match the newfound freedom and
economic power of this group. This model is very much the situation today,
where late-night transport, safe streets, quality venues and the absence of the
disreputable are demanded.

Melbin's account is worth exploring in more detail, but prior to doing
so it should not be presumed that late-night culture was already 'there', waiting
to be discovered. As the preceding section demonstrated, the illumination of
the night enabled new possibilities and opportunities that did not previously
exist. If a recognizable late-night culture began with street lights and outdoor
festivals, it has now grown into a mass of different entertainments, regulatory
frameworks and services. There has been an opening up, as it were. But the
city at night, in terms of leisure, was never entirely hidden: much of what takes
place at night now was simply not there less than a century ago. The coming of
late-night lighting or greater participation by all segments of society did not lift
the veil, so to speak, but instead functioned in conjunction with the economic
and social transformations associated with the modern city.

The American futurist Alvin Toffler, in his best-seller *Future Shock* (1971) spoke in awe of the coming 24-hour society where schools, shops, offices and factories worked around the clock in such a way that time-zones would become all but irrelevant (Toffler 1971). Melbin represents a slightly more sombre account, though, like Toffler, it has become a seminal text on the emergence of this new 24-hour society. In contrast to the work traditionally carried out on late-night culture and cities, Melbin's account is less concerned with the drinking and dancing associated with late-night culture, but rather the night as the final frontier of western civilizations. As he notes, this expansion is not uniform or necessarily consistent, and the oil crisis in the early 1970s is one example where there was a retreat of some late-night services (Melbin 1978: 7). Parkes and Thrift point to more local examples, such as the closure of news stands in Greenwich Village due to fear of crime, and the cancellation of late Mass in Boston owing to the elderly parishioners' fear of muggings (Parkes and Thrift 1980: 331). Despite these small retreats, over the past three decades the 24-hour society has continued to roll out across the developed world. While it may follow different tangents and develop according to local contexts, it would be rare now to find a city or even a town of reasonable size that entirely shuts down after 6 p.m. As well as bars and restaurants, refuse collectors, hospitals, transport providers and freight will continue to operate.

In his *The 24 Hour Society*, Leon Kreitzman considers the emerging 24-hour city from a slightly different angle, focusing instead upon the articulation of social, economic and technological drivers. For Kreitzman, the conditions of possibility for the late-night cities we now have began in the 1960s. Representing more than just the advent of the sexual revolution, the 1960s is understood as a decade of immense change in gender relations, the family, the relationship between home and work and, in a more public sense, the ways that cities functioned. He argues that as cities became 'hollowed out' due to the flight to the suburbs and the building of new towns outside the urban fringe, and as the slow slide towards the post-industrial city took hold, 'by the 1970s the thriving city centres of the 1960s were often deserted' (Kreitzman 1999: 136). More importantly, it was a time, as Kreitzman argued, when 'the great and good' no longer had the right to determine how ordinary people lived their lives (Kreitzman 1999: 136–7). This opens up a whole new world where freedom is not so much granted as taken and enjoyed in new and unique ways.

By recognizing the importance of gender, family life and youth cultures, Kreitzman understands the 24-hour city as more than simply a result of large economic shifts. The manner in which people live their lives, prepare meals, shop or work are integral to the macro-structural changes, as well as the response to those changes. This is important as it places the practical experiences of late-night users and the way they conduct their lives at the heart of a debate that could otherwise be dominated by more intangible elements.

Globalization, new technologies or national policies that may seek to deregulate or control the late-night city shape how it functions. Our lives are touched and sometimes determined by these factors that may seem out of our control or beyond our responsibility. But these factors are made redundant if we fail to consider how they operate on a local and personal level, and are therefore shaped in turn.

Working in the late-night city

Work and night-time are not generally thought of as analogous. It is the day that typically represents the mundane nature of employment, not the hours after dark. Indeed, for Schlör, day-time is about facts and work, economics and traffic (Schlör 1998: 9). These activities occur at night as well, but as he argues, *Paris de jour* does not have quite the same resonance or evoke the same kind of intrigue as the more exciting world of the city at night.

Work is certainly not anathema to the city after dark, however, and here we pursue a tangential narrative to the history of late-night leisure: the history of late-night work. Parkes and Thrift note that 'the night-time colony needs support, in much the same way as any colony does, and so an available labour supply is a necessary enabling factor' (Parkes and Thrift 1980: 329). The earliest forms of late-night work were traditionally working-class occupations such as in agriculture, mining, baking, watch services, navigation and, of course, all types of personal service. By 1900, a whole litany of late-night workers were going about their duties, from factory workers to those employed in the service sector, fashion, transport and communications, as well as those employed in the provision of essential services such as electricity, water and sewerage; jobs that, in other words, 'make possible the circulation of urban life' (Schlör 1998: 93). Marx's account of the struggle for the 'ten-hour' day highlights the extent of rampant exploitation in the UK, France and the USA in the mid-nineteenth century (Marx 1977). Such was the continuing resistance to increases in late-night employment in the twentieth century, by 1918 Brandeis and Goldmark released a report titled 'The Case Against Night Work for Women', which examined the detrimental impact of working at night (Melbin 1978: 4).

In the twentieth and twenty-first centuries, different forms of shift work have emerged: call centres, late-night gyms, store packers, etc. Workers at night are now more likely to be found in shops and call centres than factories or mines. The day-time may well continue to be understood as the time for work, but the night has now become the frontier, in Melbin's terms, for both work and the expansion of capitalism. Indeed, by the very term night-time *economy*, the late-night city is understood to be a temporal space concerned above all else with economics and the exchange of capital.

According to the Futures Foundation, approximately seven million people in the UK are now 'economically active' between 6 p.m. and 9 p.m. In Ireland, 5.4 per cent of workers finish their day's work after 1 a.m., with 13.4 per cent working throughout the night (Geiger 2007: 25). Though now dated, figures from the USA in 1976 tell a similar story, with the US Bureau of Statistics reporting that of the 75 million employed, 12 million work predominantly after nightfall, with 2.5 million of those working a shift that begins after midnight. Studies of shift-workers have been well documented, but actual figures for the number of people now employed and servicing the evening economy between 6 p.m. and 9 a.m. are not entirely clear. In London, figures suggest that up to 50,000 people work in the city's hospitality sector, but beyond the capital, it is curious that despite the claims that the evening and night-time economy are important for job creation, no figures are actually available. The temporary nature of much service sector employment may be partly to blame.

A more tangible trend that can be more easily documented is the emergence of spaces that bring together work and leisure. Network sociality and the blurring of work and play is a concept popularized by Andreas Wittel (2001). If the city at night used to be about the local and anonymous meeting up, or about purely pleasure, now it is a place for business networking and the sharing of social capital. In the period of industrialization, alcohol was seen to be antithetical to productivity. Alcohol is now central to the new network sociality where, over drinks, strangers may clinch a deal or make an important contact. Pubs and bars have long served this function, as spaces where work colleagues can come together to do informal business. In Wittel's account, however, it is people in the creative sector, typically on short contracts, who are coming together to hopefully meet their prospective colleagues or employers. More importantly, this (new) turn of events sees a variety of jobs where 'leisure' is at the heart of the new world of work. Cultural intermediaries (Nixon and du Gay 2002) such as fashion buyers, bookers and image consultants conduct their work in pubs and bars late into the night. Wittel's discussion refers to a small segment of the working population; those with family or caring responsibilities, or those over the age of 30, may struggle to recognize themselves in his account. But, Wittel's argument suggests an entirely new practice of using the night-time city for the middle-classes. As argued here, the industrial revolution witnessed the separation of work and leisure, and a distinction between zones for each activity. By the late-twentieth century, workers have returned to the night in ever greater numbers. Wittel's argument, however, points to how the boundaries between night and day, and the spatial–temporal organization of play, work and leisure, have begun to be reconfigured according to the needs of business *and* pleasure.

From an opposing perspective, Glorieux *et al.* examine the example of Belgium. Taking the years 1966 to 1999, they recount how Belgian laws passed in 1987 allowed businesses to relax late-night and Sunday working: 'This changing legislation has frequently been implicated as the starting point for a 24-hour society and economy, in which "round-the-clock" production and consumption are increasingly considered as normal practices' (Glorieux *et al.* 2008: 64). Though they found no significant increase in the number of late-night workers, this challenge to the notion of the 24-hour city and 24-hour productivity should be read in the light of the expansion of the service sector, which already functioned outside normal business hours, and the noted failure of corporate and government services to respond to changing work–life patterns.

Andrew Barr's *Drink: A Social History* (1998) provides another way of thinking about these changing patterns of late-night activity. Barr's account is less about the expansion of work into the night, but more a reminder that our use of time is not ahistorical or universal. As an example, he recounts how breakfast, dinner and the eighteenth-century fashion for lunch occurred at different points in the day for different people. Parisian customs of the nineteenth century, for example, saw different social classes sitting to dine at progressively later times, with the artisan dining at 2 p.m. and 'rich bachelors' not before 6 p.m. (Schivelbusch 1988: 140). Echoing Amin and Thrift's argument about the ever-changing rhythms of urban life (Amin and Thrift 2002), these examples demonstrate that night and day are not entirely stable concepts. Activities and conduct considered appropriate for one time can actually just as easily shift, whether that be due to economic and cultural changes, or the whims of fashion. The late-night city is irrefutably different to the city during the day, and as this book attests, there are significant variations around leisure, access, exclusion and economic rationale. Nonetheless, while these current patterns of usage may seem thoroughly natural, the 'right' time to work, eat or socialize is historically and geographically contingent, and likely to change.

As can be seen, work, leisure and a new form of work/leisure as discussed by Wittel have been pushed further into the night. Patterns of activity have also been shown to vary depending upon a confluence of technology, local custom and the economy. Context is central to this argument, and some of the trends discussed, such as the fully operational late-night city, may be more exclusive to London or Moscow than to Tamworth or Basingstoke. For Kreitzman, this 'new' 24-hour society is understood within the context of globalization – with all the vagaries and geographical inequities that entails. He argues that globalization has been a key factor behind the growth of the late-night city. We are not referring here to the internationalization of music or some alcohol brands – as important as these are – but rather how modern capitalism necessitates a fully connected and 24-hour economy. Finance, specifically global finance, is one such example. For the city to operate as a global centre it relies upon technologies and industries connected with other global hubs. In turn, for those employed in the late-night city, fundamental shifts are required in how they live their lives, partake of their evening meal, travel to work or organize their social life.

As noted by Glorieux *et al.* (2008), the 24-hour city is typically understood according to broad macro-processes and the inter-connected global city must function at all hours in order to meet deadlines and communicate with other global players at suitable times. New technologies have been developed to support this, resulting in the familiar, albeit overblown, notion that space and time have become annihilated. Parkes and Thrift point to airports and rail and bus terminals as emblematic of the new hyper-connected city, and as feeding the requirements of a 24-hour economy. They also point to the more 'obvious example' of conducting international telephone calls as evidence of a global infrastructure that functions incessantly (Parkes and Thrift 1980: 329). This 'obvious example', some 25 years after Parkes and Thrift were writing, is now a familiar and common feature of 24-hour connectivity. The call centre is but one extension of this, whether it entails staff cold-calling locally or abroad, or the cleaners tidying up for a new day/night of activity (Allen and Pryke 1994: 467–71).

Cleaners, transport providers and other support staff are a reminder of the tensions inherent within this global 24-hour city. As Massey notes, 'much of life for many people, even in the heart of the First World, still consists of waiting in a bus-shelter with your shopping for a bus that never comes' (Massey 1998: 163). In towns outside of the major global hubs, those buses may not even run past 6 p.m. Indeed, as Massey warns, there is a tendency to over-estimate the speed and connectivity associated with globalization. This is not to dispute that the world we live in today has not changed, and that the impact these changes have had on the late-night city are immense. It is possible to do things now that were not possible several decades ago, such as shopping or banking online at all hours. But, as Lash and Urry (1994) have also highlighted, the global economy is exceedingly fractured and disorderly. It would be fair to say that, for many, the interconnected late-night city is not so much about late-night traders or communication specialists, but cleaners, lorry drivers and call-centre workers.

New forms and spaces of sociality

Globalization, new forms of work, lighting and the expansion of new forms of leisure have all contributed to most recent extension of the late-night city. Fear or pleasure may remain at the centre of how the night is perceived, but other activities, and other discourses, have come to frame the ways the night-time city is understood and imagined. One final component is explored in this chapter, and that is the effect of different household structures. Just as the other technologies and practices discussed earlier have altered the way the city at night has developed and continues to evolve, different household structures impact upon late-night activity.

Earlier, we discussed how urbanism was at the heart of the revolution of US forms of leisure. As the USA's economy shifted from agriculture to industry in the nineteenth century, Somers argues that scores of young women and men moved into the rapidly growing city to take advantage of new employment opportunities. These people signified not only a new type of economy and work practice, however, but also a new form of sociality and urban living. The bars and theatres that developed were akin to melting pots where people ordinarily separated by class or social *mores* could come together. Somers places these venues within the context of new forms of community. Typically young and single, those that came to the burgeoning cities lived alone in boarding houses and hostels and sought out new forms of recreation and intimacy. Away from the traditional family structure and from the homogeneous world of rural communities, the city offered an immense array of opportunities to meet and mingle with others, be those strangers, new friends or lovers. In turn, new forms of sociality came to exist, and more importantly for the argument here, new social spaces where such meetings could take place.

This is, of course, a familiar narrative of city life. As Bauman notes, whatever the changes to the physical form, economy and way of life in cities, 'one feature has remained constant: cities are spaces where *strangers* stay and move in close proximity to one another' (Bauman 2007: 85). Britain's industrialized urban centres emerged much earlier than those in the USA, but a comparable theme can be seen in how the city is often represented as a cornucopia of new pleasures and opportunities awaiting the newcomer. Single people, whether they are fresh from education or from smaller towns, have long been seen as the vanguard of city life, and are typically represented as driving the night-time city, its shape, character and the leisure opportunities it offers.

A similar narrative is at work today, whereby as a result of single-occupancy households, the expansion of education, the expanding service sector, the urban renaissance and a reconfiguration of familial relations, cities are the hub of leisure and new forms of social interaction. No longer reliant on traditional family networks, there is today a new, highly individualized workforce, unencumbered by the traditional binds of community, family and social expectation. Though contested, this thesis has been popularized by writers such as Anthony Giddens (1998), Zygmunt Bauman (2007), Ulrich and Elisabeth Beck (1992; Beck and Beck-Gernsheim 2001) and Richard Sennett (1999). Each takes a decidedly different approach to early twenty-first-century life, but a common argument is that modern life expects entrepreneurial individuals to create their own biographies, with the city functioning as a vast backdrop in which to explore newfound economic and social opportunities through new forms of 'imagined or aesthetic communities' (Lange 2006: 153). Younger generations today are apparently far more flexible in their work life, more mobile and less likely to be encumbered by marriage or child-rearing responsibilities. The number of people attending universities has also, in the UK at least, increased substantially, and though this may not lead to greater opportunities post-graduation, and the same old exclusions based upon geography, race, class or gender continue to operate, this has, for a window of time at least, resulted in a substantial change to the development of late-night economies in university towns. For the purposes of the discussion here, these factors, amongst others, impact on the way the contemporary nocturnal city has developed. It has changed the type of entertainment offered and the spaces in which leisure and sociality occur.

The growth of single households is one such manifestation. As of 2001, in England and Wales there were 21,660,475 households, with 30 per cent of these (6.5 million) being single occupancy. This represents an increase from 26.3 per cent in just over a decade (Office of National Statistics 2003), and the total is expected to soon grow to 40 per cent. This translates to almost two-fifths of all households being single occupancy in the coming decades. In Scotland, the growth of single-occupancy households is also high, and projected to increase by as much as 35 per cent from now until 2031, with growth forecast to be centred in Orkney, West Lothian and Edinburgh (BBC News 2008). Taking a longer time frame, the number of people living alone in Britain increased from three million in 1971 to seven million in 2005. As a corresponding trend, in England in 2005, 58 per cent of men and 39 per cent of women aged between 20 and 24 remained at home with their parents. This represents an increase of approximately 8 per cent since 1991 (Office of National Statistics 2007a).

Single life, whether living alone or with one's parents well into adulthood, was once the preserve of bachelors, divorcees or widowers, but it is now an accepted and common feature of urban life. Rather than being an object of pity, the single person represents a newfound economic and social freedom, and, supposedly, an active urban lifestyle. Clearly, there are benefits to this. A society where women are no longer economically dependent upon men, are free to chose if and when to have children, and where both sexes are less likely to be tied to living in a single location all their life, is vastly different to the stereotype of the 1950s' housewife, trapped in suburbia, limited in her job opportunities and tied to an unhappy marriage. Men, too, may now make more active decisions in their life, and with the average age of their first marriage being 28, there is an equally large number of men as there are women who have a greater degree of social, economic and possibly sexual freedom born out of the trend towards single life.

What this has to do with the gestation of the modern late-night city is complex. Just a decade ago, Justin Worsely from the Henley Centre claimed that the increasing number of single households would result in:

> people ... using out-of-house activities more for socialising. Restaurants, pubs and cinemas will be their social base, especially as the entry costs for a nice restaurant tumble down in relation to a fast-food joint ... One thing you'll see is a growth in suburban coffee bars, moving from the inner city.
>
> (Rowan 1998: 13)

It is important, nonetheless, not to over-estimate the true picture of single life today. The lives of single women depicted in US dramas such as *Sex and the City* or Britain's own *Bridget Jones' Diary* appear to speak directly of a new-found freedom. And yet, the lives of single people in Britain today are vastly different to those lived in televisual households. Before one becomes too breathless at the economic and cultural power of single men and women, of the 6.5 million single households referred to above, almost half (3.1 million) are made up of pensioners. The vast majority of these pensioners, three-quarters in fact, are women living alone. Single-person households are, moreover, least likely to have central heating or their own bathing and toilet facilities. This is equal to 383,000 pensioners and over 430,000 non-pensioners in England and Wales today without central heating or their own bathroom (Office of National Statistics 2003).

Single-person households may represent one factor behind the growth of contemporary late-night culture and the emergence of new spaces for social networking and leisure. Promotional material for the urban renaissance will certainly sing the virtue of inner-city lifestyles to young single folk (Hoskins and Tallon 2004). Care needs to be taken, however, in order to not over-estimate this demographic in such a way that the struggles of many single households are ignored. Moreover, as Bennett and Dixon suggest: 'City centres may become increasingly dominated by people living alone, but there is little evidence that the increased popularity of city centre living is creating benefits for deprived inner urban areas' (Bennett and Dixon 2006: 4).

Just as there is a clear need in terms of basic services such as central heating for single people, there are, for the discussion at hand, 'wants' that may not be met by current policies, either. Though the highest number of single households are elderly women, the fastest-growing demographic are 25–44-year-olds, with young men representing the highest growth (Bennett and Dixon 2006: 3). For older people, living alone is more likely to be the result of bereavement or divorce, but for this rapidly growing younger group it tends to be a choice (Bennett and Dixon 2006: 9). Indeed, as Nathan and Unwin

(2005) have noted, the urban renaissance is dominated by young people, many choosing to live alone. More importantly, these households are largely confined to inner-city areas. In the central London boroughs of Camden, Islington and Westminster, for example, 40 per cent of households are now single. Comparable figures are found in Brighton and Glasgow (Bennett and Dixon 2006: 11). There is a debate concerning the extent to which single households contribute to their local community, or foster community relations in a recognizable form (Bennett and Dixon 2006). That they influence the creation of new spaces for entertainment, retail and service provision, however, is less contested.

If young men comprise the fastest-growing number of singles, and if singles can be seen as a key factor driving the creation of new spaces for sociality, a corresponding factor is the growing number of students in higher education. Britain's educational expansion has occurred over the past two decades, seemingly in tandem with the rise of the night-time economy. The nocturnal city has not grown solely in response to student culture, but it has become central to how student culture is represented, lived and imagined. Many of Britain's most vibrant night-time economies are in fact centred in or alongside universities and colleges (Chatterton and Hollands 2003). Binge drinking amongst students in the USA has caused alarm in recent decades, but British students are even more likely to entertain themselves in alcohol-related venues. The difference in legal drinking age can be used to partially account for the fact that visiting pubs is a more common activity in the UK than the USA, where sport and visiting family or friends account for the two most popular activities. In Britain, approximately seven hours are spent per week in pubs (Chatterton and Hollands 2003: 130), while US students spend almost twice that actively engaging in sports (Chatterton and Hollands 2003: 130).

This rise of student numbers and growing alcohol consumption should not necessarily be seen as interchangeable. York's infamous Micklegate Run, where students would drink a pint in each of the street's bars, dates back to well before the current evening economy or growth of the education sector. Other drinking games and customs have long been popular with student culture, and the drunkenness of fraternity houses has been well documented. However, with, on average, £1,000 (US$1,534) per year spent on entertainment, students comprise a considerable market for late-night businesses and therefore play a considerable role in the ways the late-night city has developed.

The situation for women is slightly different. Recent years have seen a broadening in the representation of single women and late-night culture. Far from Schlör's 'woman of questionable morals' who ventures out after dark, women today may well be consumers, DJs, cultural producers, service providers or security staff (Hobbs et al. 2007). A body of work has also emerged that specifically examines women's experiences in rave and club cultures (Hutton 2006; Pini 2001; Thornton 1995). Despite these more productive accounts, there is an equally strong corresponding image featured heavily in the tabloid press of the young 'irresponsible' woman, drunk in a gutter. As is explored elsewhere, even today, women attending hen parties are routinely figured as 'slags' or 'cheap', as if their mere presence on the streets at night signifies a desire for sex. Men's pleasures in the city at night, and their projections, will often curtail women's similar freedoms to wander and delight in the city after dark (Schlör 1998: 168), and where women are free to wander and explore the late-night city, much of this freedom entails a freedom from having to justify their presence. Pejorative images may do little to detract other women venturing out, however, and may even be seen as little more than attempts to police women's late-night excursions. After all, though not to the same extent as today, women have always been participants in late-night cities. The perception that women have only recently left the home to partake in nightlife ignores the vast army of women who worked at night, as well those more privileged women who attended operas, pleasure gardens or the theatre.

The post-1960s increase in women participating in the paid workforce is a key driver in the growth of the contemporary late-night city. In fact, the modern increase in the night-time economy has occurred at the same time, and rather than being a side-issue for analysis, or assuming women have been absent from the contemporary late-night city, it appears they have been central, if not active drivers, of almost every major new initiative over the past four decades in the 24-hour city. The emergence of chain bars, alternative cultures such as rave parties, the modern superclubs and the demise of traditional men's boozers, late-night shopping and late-night service-sector workers; women have been central to each and every one as both producers and consumers.

This is not to say the foundational issues are not still there. Women's fears of venturing out at night are unfortunately still current (Pain 2001). And while, as is now widely known, sexual violence is more likely to occur in the home than on the streets, the hen party example discussed elsewhere demonstrates that women continue to have to negotiate outdated perceptions of appropriate feminine behaviour (Day et al. 2004). As a further example, the increasing number of women in paid employment is often used to explain women's increase in alcohol consumption. The perception that women drink more today simply because they 'work' is deeply problematic. Not only does it ignore the fact that women have always worked, it tends to obscure far more relevant details such as later marrying age, delayed childbirth, more suitable venues and the acceptability of women participating in late-night culture – as both consumers and producers.

A far more useful thread to explore is how, as the number of women venturing out has increased, specific spaces and forms of connection have also opened up. Chatterton and Hollands examine 'feminised' establishments with large windows and extensive wine lists as evidence of a new, more welcoming atmosphere for women (Chatterton and Hollands 2003: 149). The cultures Thornton (1995), Pini (2001) or Hutton (2006) explore emerge out of a more complex web of shifting economic and social relations: changing attitudes to leisure, higher levels of study, living in multiple-occupancy households without the constrictions of domestic life, and a later age in marriage and childcare responsibilities. In 2006, women giving birth for the first time were, on average, aged 30 in England and Wales, compared with 24 in 1971 (Office of National Statistics 2008b). A later age in marriage for women, from 24 in 1971 to 30 in 2006, is also a factor that creates the conditions of possibility for a much more accessible late-night culture for both sexes.

In summary, there are key drivers behind the growth of the night-time city, as well as new consumer markets. These groups have facilitated the growth of different forms of late-night activity, from rowdy and youth-focused student bars, to the requirement for more domesticated social spaces for single people to meet and network, as well as new types of alcohol-related establishments for women that are vastly different from the traditional men's pubs. In Chapter 9 we return to some of these themes and examine how ordinary consumers experience the nocturnal city. The role of gender and different forms of entertainment are also explored in Chapter 5 in relation to binge drinking. Before returning to these themes, we turn to the revitalization of British cities in the 1980s, and the emergence of the contemporary late-night city. These are the spaces that have resulted from the many changes discussed above. The introduction of lighting, the growth of the late-night service sector and changing patterns of work are some of the key changes identified that have, with different intensities in different locations, impacted on how the modern night-time city has developed.

Chapter 3

Visions of the
night-time city

the wind of the so called **urban renaissance** came over the cities...
(Oosterman 1994: 123; bold in original)

Introduction

In the early 1980s in Britain, town and city centres went through difficult times.
At the beginning of the decade a freshly elected Conservative government
intent on change facilitated the re-structuring and de-regulation of British industry and commerce. 'Old' industries were allowed to go bankrupt or to re-locate
their plants overseas. Traditional commercial institutions such as banks and
building societies re-structured their organizations and methods of working.
Unemployment rose and many town centres suffered from reduced demand in
the shops and pubs and general decline in facilities and services.

Planning policies derived from post-war modernism, such as zoning,
the creation of monofunctional areas and the primacy of the car, achieved frui-
tion and a new urban landscape gradually emerged: ring roads, high-speed
motorway links and out-of-town 'big shed' developments. As prosperity
increased in the last half of the decade the residential and commercial property
markets boomed. House prices rose rapidly and unemployment declined. Town
and city centres still struggled, despite an increase in prosperity. Industrial re-
structuring left industrial buildings located near to town centres, such as warehouses and factories, derelict. Although the financial services industry expanded,
it also re-structured as 'new' technologies of telecommunications were adopted
(Graham and Marvin 1996). Staff were moved to 'back offices' on the outskirts

of metropolitan agglomerations (Castells 2000), leaving handsome examples of banking halls and insurance offices empty. Changes in entertainment were also adding to the experience of redundancy. Although colour television had been available since 1967, ownership had been increasing and the appearance of videos and other technologies resulted in widespread prophecies about the 'death' of the cinema. In actual fact, once entertainment providers realized the potential for attracting audiences to contemporary comfortable auditoria, this trend was reversed. Nevertheless, new cinemas tended to be established in new multiplexes at the edge of town, rather than in existing, and often architecturally distinctive, historic buildings in the town centre.

It was against this background that fresh approaches to town centre regeneration were formulated. '24-hour cities', the 'creative city' and cultural quarters were each policy responses to decline and a fear of a 'lost' urbanity. In the big cities, outside of London, Glasgow and Edinburgh, the residential population had all but dispersed to the suburbs. There was a real danger that the centres of once-powerful metropolitan cities would become hollow shells. Even in relatively prosperous market towns, living 'above the shop' had become stigmatized and new housing was constructed on the fringes. The 'retail revolution', that is the process in which small, local shops are being supplanted by major corporate chains, was just starting as new hypermarkets were constructed on the bypasses and ring roads. While these more prosperous locations hosted shop fronts of building societies and estate agents rather than derelict buildings, similar concerns were expressed at a loss of 'life' and vitality. The spectre of a North American style of urbanism, car-based, home-centred and low-density, stimulated consultants and professionals to re-discover the qualities of European urbanism.

The '24-hour city'

Ironically, it was not a European theorist who inspired 24-hour city policies, but a North American. Jane Jacobs' classic work, *The Death and Life of Great American Cities* (Jacobs 1961), is well known. Although it had been widely taught to generations of architects and town planners in the UK since its first publication, it somehow failed to reach policy and practice. Jacobs' message was not easy to implement. Basing her ideas on ethnographic observation in New York and Boston, she provided a swingeing critique of comprehensive redevelopment, which she dubbed as 'catastrophic' and argued instead for incremental change. She was a passionate advocate for dense, mixed-use neighbourhoods, which she demonstrated were not the disorganized slums derided by City Beautiful and Garden City enthusiasts, but were the product of 'organized complexity'.

The chapters that inspired 24-hour city and, indeed, contemporary urban renaissance thinking, were those that analysed the 'uses of sidewalks' and the 'ballet of street life'. Here she drew on her own observations to explain how busy city streets provided safety, comfort, variety and interest from early morning until late at night. Presaging the work of Sennett, she pointed out that cities are composed of strangers who observe a certain social and emotional distance in order to coexist in close proximity to each other (Sennett 1996). Combating the charge that cities are anonymous and uncaring places, Jacobs instead painted a picture of her own neighbourhood. Here, she observed that the shopkeepers and little business owners 'looked out' for each other and for the children of local residents. They would tackle petty crime, incivility and potentially threatening situations through direct interventions. Their guardianship of the street would extend throughout their hours of trade and commerce. In the type of street that Jacobs described, there was always some business that would be open for 18 hours of the day, from an early-morning café or delicatessen to a late-night bar or a graphic designer rushing to meet an order.

Jacobs' observations were firmly based in the 1950s, when street neighbourhoods were relatively homogeneous in their social mix (Berman 1983) and small businesses could thrive. Although Jacobs has been criticized for romanticism in her depiction of street guardians, she was not naïve about the tendency of businesses to cluster. As Hadfield (2006) notes, Jacobs commented on the problems caused by too many restaurants or bars clustering in one place and suggested regulatory measures to prevent that. These issues did not prevent urban theorists in the 1980s from drawing on Jacobs' ideas and injecting them into the context of late-1980s Britain.

A report produced by a consultancy then called Comedia was particularly influential (Comedia 1991). Based on an observational study of town and city centres in Great Britain, *Out of Hours* focused on the hours after the shops closed, typically 5 p.m. or 5.30 p.m. The report noted the desolate nature of many town centres: pedestrianized precincts with nobody in them; pubs that seemed to exclude outsiders; and the presence of boisterous young males whose behaviour scared away older people and families. The Comedia authors proposed a number of initiatives in different policy areas to breathe life back into town centres. They argued that it was essential to introduce residential uses back into town centres, and to provide the presence of people who would create a demand for a variety of different facilities, from convenience stores to entertainment. Furthermore, a strong residential population would provide a critical mass of people who would be on the streets at different times of the evening and night and, through their everyday presence, enhance feelings of safety.

In addition to challenging the monofunctional nature of town centres, Comedia also proposed measures to make them more inviting for visitors and tourists. Their vision was of the type of town centre that is found in mainland Europe, particularly in the cities of southern Europe. British tourists visiting seaside resorts in Italy, Spain and Greece in the 1980s reported their appreciation of the way in which whole families could be seen out together in a group until nine or ten o'clock at night. Young children were welcomed into restaurants and it was possible for groups of people of different ages to sit in cafés for extended periods of time, drinking alcohol and soft drinks in a relaxed manner. This contrasted sharply with British pub protocol in the 1980s, where venues were dominated by alcohol, standing up, round-buying and following various unwritten customs and rules (Fox 2005).

The proposals for instituting a culture change were far-ranging and included environmental improvements to the public realm and making streets and public places more attractive for sitting out and walking in. These environmental improvements drew on a synthesis of urban design analysis and prescription developed at Oxford Polytechnic in the 1980s (Alcock *et al.* 1988). These included not only townscape and streetscape improvements such as paving, but also more 'structural' changes such as providing continuous street frontages and an active engagement between the pavement and the activities behind the building façade. Livelier streets would be created by permitting café tables to be set out at night and encouraging pub owners to install plain glass windows rather than the traditional frosted glass. Improvements to lighting, the use of flags and banners would enhance feelings of safety and provide information and excitement.

The authors challenged conventional notions of transport provision and infrastructure. They suggested that the practice of pedestrianizing town centres was counter-productive at night, because when footfall was lower, near-empty streets were forbidding and the presence of drunken youths seemed more threatening than perhaps their behaviour would warrant. Reconsidering pedestrianization schemes or reversing them at night to permit traffic flow would help to resolve this issue. Improving public transport at night was

3.1
**The Ramblas, Barcelona at
10 p.m. This relaxed scene
typifies a more inclusive
night-time culture.**
Source: photograph by
author.

an obvious demand, but recognizing the difficulties in achieving this, the team also suggested improving evening and night-time car parking facilities, including routes to and from the car parking areas.

A critique was also mounted of British licensing legislation, which at that point had only recently been liberalized to include afternoon opening, and in most centres still operated with the strict terminal hours of 10.30 p.m. for pubs outside London and 2 a.m. for nightclubs. Violence was attributed to the 'macho' culture attached to pub drinking, with its ritual of buying 'rounds' and drinking fast at the end of the evening to finish by the terminal hour. Suggestions were made that changes could be achieved by attracting more families out in the evenings and making pubs more 'women-friendly' through conversion to café bars and wine bars. An extension of licensing hours later into the night would not only reduce violence but would encourage more footfall and life into the early hours of the morning.

The Comedia authors each elaborated and extended these ideas through books, articles and conference papers in the years to follow. Bianchini brought forward experience gained from Italy, where the City Council in Rome had run a series of summer night-time events in the period 1977–1985 (Bianchini 1995). This initiative had entailed cultural activities sponsored on four different sites in the city with cheap and frequent bus services between them. As a consequence the whole city became infused with cultural events such as films, circus shows, jazz, classical music and dancing. Both Bianchini and Montgomery proposed a comparable programme of 'cultural animation' events to enliven the evening and night-time, suggesting that towns and cities hold free festivals and events in the evening and at night to attract audiences back into the centre from the suburbs and outlying settlements (Montgomery 1998).

Worpole, one of the *Out of Hours* authors, pursued similar themes in his book *Towns for People* (1992). His later work also recognized the problem of drunken anti-social behaviour, and subsequently placed great importance on reducing public drunkenness in order to create more convivial town centres. Drawing on a Home Office study, Worpole attributed concentrations of violence in town centres to the commonality of pub closing times, and called for more liberal licensing laws to combat this. Despite this apparently negative comment on the state of town centres, the book argued that British town centres actually contained a great deal of life and activity in the evening hours, with many community halls in the back streets enjoying full bookings from voluntary and amateur cultural and social groups. The theme of the underlying 'richness' of cities has permeated later works and Worpole has championed an awareness of the value of establishments such as public libraries as community and social institutions (Worpole and Greenhalgh 1999).

The Comedia authors highlighted the 'gap' that occurred in many town centres in the time between the shops closing and people coming out for an evening's entertainment. Some ideas were put forward to revitalize this 'dead' period, such as extending shop opening hours. The rationale behind this was to retain town centre workers in the centre through to the traditional time for 'going out'. The same rationale lay behind the idea for extending the practice of 'happy hours', a period when pubs and other venues would offer drinks at reduced prices. As will be seen in subsequent chapters, this practice was to have negative unforeseen effects in terms of public drunkenness and alcohol abuse (Miles 2005).

The concept of the '24-hour city' comprised a series of propositions to revitalize town and city centres. It encompassed more than an idea of 24-hour drinking. As Heath recounts, the ideas were swiftly taken forward into policy by a number of British cities who were looking for ways to revive city centres that would not require large redevelopment projects (Heath 1997). Cities such as Leeds, Cardiff, Nottingham and Manchester adopted the ideas. Even Westminster in central London had suffered during the economic downturn and welcomed a way of boosting tourism and entertainment (Westminster City Council 1997). One of the first experiments in extending licensing hours was carried out in Westminster, where many clubs were allowed to open until 3 a.m. and 40 were allowed to remain open until 8 a.m. on Saturday and Sunday mornings (Jolly 1994).

In 2003 a major conference was hosted by Manchester Metropolitan University to critically review 24-hour city initiatives and policies (Lovatt et al. 1994). Montgomery was a key contributor and organizer. He developed the theme of the night-time city as a contributor to the local economy and pointed out that the economic impact of boosting evening entertainment uses extended further than an increase in employment in pubs, bars and clubs. The effects could be felt through various supply chains – for example, in lighting hire, electrical equipment such as PA systems, legal and financial services (Montgomery 1994). Montgomery highlighted the multiplier effects of the evening economy, that is how a cinema or theatre could help support smaller eating and drinking establishments because customers would go to have a drink or a meal before or after the show. Jan Oosterman, a planner from Holland, where cities such as Amsterdam and Utrecht had been promoting their evening economies for years previously, confirmed this view (Oosterman 1994).

Oosterman also emphasized the significance of the pavement café and its role in providing vitality to the evening street scene. The street café, he argued, was a place where the heterogeneity of city life could be observed without threat. It provides a vantage point, where one could see and be seen and where accidental encounters between acquaintances could be facilitated, more easily than in the seclusion of a pub bar. This was a theme taken up by a number of speakers. In later years, the notion of the 'cappuccino' society has come to stand as a kind of shorthand for an image of a civilized, predominantly middle-class European urbanity (Atkinson 2003; Montgomery 1997).

The concept of the 24-hour city could therefore convey a multiplicity of meanings. It could stand for pavement cafés, for an increase in cultural and entertainment uses, for a relaxed approach to liquor licensing, for city centre living and for good urban design in the public realm combined with night-time events. The term itself became a slogan that could be applied to cities as varied as Swansea in South Wales and Bolton near Manchester. As an idea about regeneration, it dovetailed into the other dominant theme of the 1990s, which was cultural regeneration.

Cultural regeneration: cultural quarters and 'creative cities'

The 'cultural turn' to urban policy-making has received much academic attention (see, for example, Evans 2001; Miles and Paddison 2005). It is not the intention of this chapter to provide a comprehensive view of cultural policies or, indeed, of the burgeoning literature on cultural quarters (Bell and Jayne 2004). Rather, the intention is to point to aspects of the cultural turn that have assisted in boosting activities in the evening and at night. This is complicated because the temporal nature of much cultural and creative activity has received relatively little attention in the many and varied debates on the topic. Cultural and creative activities blur and re-cast temporal boundaries, and few can be said to be day-time or night-time only. For example, nightclubs and late-night takeaways are the predominant form of late-night activity in the UK, whereas in the USA 24-hour diners are a more commonly found feature in the big cities. As was discussed in Chapter 2, the creative industries often demand different working hours to the UK standard day-time.

As McCarthy notes, each major town or city centre now has a 'cultural quarter', some formed 'organically', others more formally planned or 'designated' (McCarthy 2005). Myerscough's (1988) groundbreaking study of the economic importance of the arts highlighted the synergies between explicitly cultural consumption and 'hospitality'. He demonstrated that for every trip to the theatre or the opera, a multiplier effect was produced by members of the audience also going to bars and restaurants nearby for a meal or a drink. These synergies were welcomed by policy-makers and helped to justify public investment in 'high culture', such as building new theatres and opera houses. The quantity and range of night-time venues in many major UK town and city centres has increased in recent years in the expansion of culturally led regeneration. Further examples of 'quartering' cities, such as designating micro areas as 'Chinatowns' or the 'Irish quarter' have boosted the parts of the hospitality sector that are run by owners drawn from ethnic minorities and offer speciality eating and drinking. If the definition of everyday culture is extended to a way of life, then eating and drinking can also be regarded as cultural activities (Bianchini and Ghilardi 2004).

The distinction between culture and 'hospitality' is stretched by the appearance of 'designer' bars, clubs and restaurants. Barcelona was one of the first cities in Europe to enjoy venues where the experience of everyday activities such as drinking alcohol and dancing was provided with a chic legitimacy by architect-designed interiors. A younger generation of architects, stymied by the economic downturn of the 1980s, were offered the opportunity of experimenting with their design skills to create a fresh approach. Their empathy with their (mainly) youthful clientele enabled them to merge and heighten the experience of the music and the lighting. In the British Isles, prominent architects were commissioned to design restaurants, rather than clubs and bars. David Chipperfield, for example, designed one of the first of the chain of Wagamama restaurants in London's Soho. He also designed the showrooms of a fashion shop, Joseph, located in London's West End. Academic analysis of this type of 'aestheticization' of new city spaces has pointed out that they can be understood as the founding of a new symbolic order, re-positioning cities as centres for consumption (Lash and Urry 1994; Zukin 1995). Such re-positioning has gained in importance as policy-makers review the league tables of cities as desirable locations for international business.

Returning to cultural quarters, there has been much debate about their definition and whether a cultural quarter can only be defined as such if it includes cultural production as well as consumption. Further, a somewhat fuzzy link has been made between aestheticized consumption and 'creativity'. A whole theory of a 'creative class' has been put forward by the economist Richard Florida, who argued that the most dynamic cities in the USA were those that supported a bohemian-inclined layer of workers whose values were at odds with those of corporate America (Florida 2002). The dynamism of this class lies in their ability to innovate and develop new entrepreneurial initiatives. Among their preoccupations is an interest in a vibrant nightlife. This nightlife too, is at odds with corporatism, and Florida emphasizes the value of 'cool' bars, cafés and performance venues in a historic, architecturally rich environment.

This type of nightlife is most likely to emerge from what Shorthose (2004) has termed a 'vernacular' cultural quarter, that is a cultural quarter that has emerged through synergy of different factors, such as the availability of cheap rented space in buildings with aesthetic potential and the entrepreneurial energy of a set of creative individuals who may be engaged in both production and consumption. Montgomery, who claims that Comedia and the British American Arts Association actually invented the cultural quarter, agrees with this view. Indeed, he goes on to argue that a strong cultural quarter not only possesses these characteristics but has, as a pre-requisite, a vibrant evening economy:

> Successful cultural quarters will almost certainly have a strong *evening economy* ... for where this is lacking a place can only be said to work half the time.... Indeed much of the attraction of cultural quarters is that it is possible to merge the day into the night, and formal cultural activities with less formal pursuits such as meeting friends for a meal or a drink.
>
> (2003: 297; italics in original)

The theme of nightlife as facilitating 'networking' between members of a creative class is set out in a literal manner by Florida, and permeates Montgomery's championing of cultural quarters. Both argue for policy support for sustaining this type of 'creative milieu'. It should be emphasized that the physical places that they are discussing are not the corporate entertainment districts in the downtowns of US cities (Hannigan 1998), but are districts or neighbourhoods that incorporate a mix of architecturally interesting old and new buildings, with a mix of land-uses as well as culture, creativity and entertainment, where small- and medium-sized businesses flourish. For example, Montgomery was involved as a consultant in the development of Dublin's Temple Bar as a cultural quarter in the early 1990s (Montgomery 1995). Florida also cites Temple Bar as a prime example of a creative milieu, providing a space where creative producers can pursue an urban lifestyle where 'work, ideas and friendships' can be played out in 'coffee houses, bars, restaurants, clubs, venues, galleries and other semi-public meeting places' (Montgomery 2003: 299). This then was the vision for the revitalization of town centres. It proved attractive both to city dwellers and, as will be seen, to visitors and tourists. Before going on to examine how this vision came under pressure from changes in the alcohol and entertainment industries, it is important to note how many inner-city urban areas across the globe underwent a renaissance in the last two decades of the twentieth century.

Urban renaissance

In Britain, this 'renaissance' anticipated policy. In the main it was driven by gentrification in the residential property market, combined with changes in social relations. Authors such as Butler and Hamnett have charted the process and nature of gentrification in inner London, its domination by the professional and managerial middle-classes and its spread through formerly working-class neighbourhoods such as, for example, Barnsbury, Stoke Newington and Clapham (Butler 2003; Hamnett 2002). This phenomenon was not confined to London; neighbourhoods in other major cities, even those experiencing a downturn, such as Jesmond in Newcastle-upon-Tyne, followed suit. Butler's work in the 1980s demonstrated how these middle-class fractions chose to live in the inner city as a positive choice made from the point of view of lifestyle, and not as a result of a lack of alternatives. While Butler's surveys demonstrate the continuity of middle-class values and aspirations amongst gentrifiers, others have highlighted the relative freedoms offered by such areas in forging new styles of gender relations. Appleton (1995) has suggested that a new 'gender regime' existed in certain neighbourhoods in the 1990s, where single parenthood and lesbian-headed households were accepted, a movement in advance of a more widespread change in social attitudes. The influx of higher-earning residents into these areas provided an expanding customer base for food and drink outlets of all kinds, in addition to other forms of consumption and cultural activities.

The expansion of further and higher education in the UK in the 1990s, as discussed earlier, has given a further boost to consumption. To reiterate briefly, successive governments since 1989 have adopted a strategy to encourage a higher skill base amongst the UK workforce in order to locate the UK as a knowledge-intensive nation within the global economy. In the decade 1991/1992 to 2001/2002, the total number of students in higher education institutions in the UK increased by over three-quarters, from 1.2 million to 2.01 million.

Planning policy in the UK has been put into a position where it has had to 'catch up' with these shifts in urban movements. Until the mid-1990s, policy was still based on the zoning of separate land-uses as a guiding principle (Coupland 1997). Although educationalists were teaching the merits of mixed-use development in their urban design and planning courses, and certain members of the built-environment professions were pressing for it, mixed-use development and the encouragement of residential uses in city centres only became a core plank of central government policy in 1999, with the publication of the Urban Task Force Report (Urban Task Force 1999). The government gave the Urban Task Force the mission to 'recommend practical solutions to bring people back into our towns, cities and neighbourhoods'.

Bringing people back to live in the city centre has been part of UK government policy since 1995, and – to a degree – there has been some success. Estimates for expansion vary, and although the figures are not dramatic, they suggest a significant increase. For example, the resident population in the core areas of Liverpool grew from 10,000 in 1991 to 13,500 in 2001, with the figures for a similar period in Manchester as 3,500 and 10,000 (Nathan and Urwin 2005: 12). The expansion of city centre living in Leeds has also been well documented and monitored since 2003 (Unsworth 2007). There were 1,800 completed new flats in Leeds in 2003, which increased to 5,700 by the first quarter of 2007. A further 3,812 were under construction and 5,600 had been granted planning permission. These figures exclude developments targeted at students. These statistics provide some idea of the scale of residential growth in different UK city centres, with Liverpool representing the lower end of the spectrum and Manchester and Leeds the higher.

Two surveys, one of Manchester, Leeds and Dundee, the other of Leeds, also provide information on the type of people coming to live in the newly constructed city centre flats. The demographic of the population moving in is predominately young. In Leeds 61 per cent of the survey sample were 30 years old or younger, with 18 per cent aged 41 or over (Unsworth 2007: 10). This profile is repeated in Manchester and Liverpool (Nathan and Urwin 2005: 20). Whereas the stereotype of the new city centre dwellers is that they are affluent professionals, 39 per cent of the sample in Leeds earned £35,000 or less, which puts them into the middle-to-lower earning range. In Leeds 56 per cent of those surveyed were renting their flats, with the overwhelming major-ity, 80 per cent, living in furnished flats acquired through a letting company. For young people, the rents of city centre flats are such that a flat share is advanta-geous, and 10 per cent of the Leeds sample was students. The study of Man-chester, Leeds and Dundee shows a similar tenure pattern for the newly built blocks, but with a higher proportion of social rented properties than in Leeds. These three cities also have higher proportions of students or people living stu-dent-like lifestyles living in their city centres; up to 40 per cent in Manchester.

Despite the pervasive myth of city centres being dominated by highly motivated young people eagerly embracing urban lifestyles, in each of these cities there is a significant proportion of low-income residents. In Man-chester, for example, 20 per cent of new city centre residents are in hardship and their access to a consumerist lifestyle is severely limited (Nathan and Urwin 2005). This is borne out by considering Hoxton, in London's inner east, where, despite its reputation for a vibrant night-time economy, there are a large number of council tenants. Figures from Hackney's internal audit of housing demonstrates that in 2001, and of a total of 86,040 dwellings, 30.7 per cent were rented from the council with a further 20 per cent rented from a social landlord. A further 1.5 per cent were shared ownership. More specifically, the ward with the greatest number of social and council tenants in Hackney is the late-night hotspot of Hoxton, at approximately 65 per cent (Hackney Council 2007: 80).

There is therefore a divergence of lifestyles within the incoming population. The suggestion that new city centre residents include a significant number of divorcees, or 'empty nesters', that is, older people whose children have grown up and left home, does not appear to be entirely accurate. The other stereotype, that of the affluent, highly paid professional or professional couple, may be more accurate, but it does not apply to Dundee, where employment levels and wages are low.

The Urban Task Force's recommendations validated an approach to mixed use, stating that a key principle of urban design was 'the encouragement of a diversity of activity and uses at different levels', that is, within the building, block and the street as well as the neighbourhood. Planning guidance for town centres at that time also noted the adverse impacts of an over-concentration of bars and clubs with regard to litter, noise and anti-social behaviour (Department of the Environment 1996). It suggested that some larger centres should consider a specialized leisure or entertainment quarter. By contrast, the Task Force suggested that conflicts could be overcome through 'careful planning, design and siting', although it did not explicitly refer to night-time uses in this statement. In fact, evening and night-time uses are only fleetingly referred to in this report. A checklist of design issues includes '24-hour use' at the end of a long list under the heading 'Urban form and public space' (Urban Task Force 1999). The Calls and Riverside district in Leeds is cited as an attractive newly regenerated mixed-use area that incorporates housing, entertainment, cultural, creative and other uses. The Task Force had visited Barcelona as part of its evidence-gathering activity, and the Mayor of Barcelona wrote the foreword to the report. Although the Task Force concentrated on the quality of the new interventions into the urban fabric, it was Barcelona's quality as a compact European city that impressed them. Barcelona has no central business district, or other specialized neighbourhoods. The Task Force recommended a 'pyramid' of densities, and its approval of Barcelona and a recommendation to prevent city councils from setting an upper residential limit to density gives an implicit template of the 'Barcelona model' as guidance for city centre regeneration.

The 'Barcelona model' and the vision for a vibrant cultural quarter are both attractive urban archetypes. Although they may appear to be templates that could be adopted by a wide range of cities, they are in fact based on particular assumptions about the development industry and the structure of the entertainment industry. Montgomery (2003) specifies these in stating that a successful cultural quarter is run mainly by small- and medium-sized independent enterprises.

Ley (1996) has provided a benign view of this process, drawing on experiences in Vancouver, where he describes neighbourhood gentrification as producing revitalized streets with interesting delicatessens, specialist shops and restaurants. Latham also gives a similar picture of a neighbourhood in Auckland (Latham 2003). By contrast, Smith's account of the colonization of the East Village in New York City by 'urban pioneers' discusses how these high-value new uses drove out facilities for poorer inhabitants (Smith 1996). The processes of gentrification and the spread of upmarket restaurants and bars do not necessarily go hand in hand. The growth of alcohol-related entertainment has a dynamic that is influenced by other factors. These include the property market, the structure of the alcohol and entertainment industries, planning, licensing and other legislation. It is to these that we shall turn to next.

Changes to the alcohol and entertainment industry in the UK: early beginnings

In the mid-1980s, eating and drinking outside the home in Britain had a fixed pattern. Restaurants were frequently run by small-scale independent owners, and with rising immigration there was a particular expansion in Indian and Chinese establishments. Similarly, the takeaway market was expanding, including the appearance of Greek and Turkish takeaways selling kebabs. If not consumed with a meal in a restaurant, alcohol was mainly drunk in a pub owned by large breweries. Manager publicans rented the pub from the brewery and had contracts that obliged them to sell the breweries' beer. The numbers of nightclubs and live music venues was small, and nightclubs were subject to strict controls.

The activities of the major brewers came under criticism from organizations such as CAMRA (the Campaign for Real Ale), which lobbies to protect and promote small breweries and independent operators. CAMRA looked back to the history of British brewing and regretted the loss of small independent breweries, selling a beverage that not only tasted better than much of the beer and lager produced by the large producers on an industrial scale, but which maintained a regional identity. The Conservative government in the UK was also committed to a 'bonfire of regulations' and to opening up British industry to competition. In 1989, a government report from the Monopolies and Mergers commission condemned the stranglehold that the big breweries had over beer production, with a control of 88 per cent of the market (Chatterton and Hollands 2003). The Supply of Beer Orders Act that followed in the same year prohibited brewers from owning, leasing or having an interest in more than 2,000 pubs. In these pubs, at least one guest beer had to be sold.

Brewing and pub owning were therefore separated. This provided the opportunity for new companies, 'pubcos', to take over licensed premises and to experiment with a new 'offer'. Changes to licensing assisted with this process. The Licensing Act of 1988 permitted continuous opening hours for weekdays, thereby allowing the opening of continental-style cafés between the hours of 11 a.m. and 11 p.m. in London. This was further continued by a revision of the legislation that permitted all-day Sunday opening via the Licensing (Sunday Hours) Act 1995. 'Chameleon' bars were invented, whereby the same venue could move seamlessly from being a café-bar at lunch-time, operating through to the early evening, after which the food offer would be withdrawn, the music turned up and the place would be transformed into a bar cum dance bar cum nightclub.

New clubs were emerging, stimulated by the 'rave culture' and new music of the 1980s. The history of the Hacienda club in Manchester has been recounted to great effect by Hobbs *et al.* (2003) and immortalized in the film *24 Hour Party People* (2002). This club, which achieved international fame, was set up on the fringes of Manchester city centre by a (relatively) young entrepreneur, Tony Wilson. Wilson also managed emerging bands such as Joy Division and the Happy Mondays. The club flourished, despite the business inexperience of its owner and managers. Imitators followed, making use of the atmospheric but newly redundant industrial warehouses outside the immediate city centre, spaces that had development potential but which conventional developers were oblivious to. Lovatt points out that these new bars, restaurants and clubs owed their success to a 'design ethos which was informed by public demand for new urban spaces' (Lovatt 1994: 33).

Not only did developers ignore the opportunities presented by the new youth culture, a culture that was drawing people into the city centre en route to their 'night out', but in Manchester the police and council officers positively frowned on it. Their attitudes date back to the 1960s, when clubs had been associated with, in the words of the Chief Constable, 'thieves, prostitutes and homosexuals' (Lovatt 1996). Police attitudes in the early 1990s had some justification, as the Hacienda did fall victim to a gang shoot-out over drugs. It was temporarily closed down by the police, but was allowed to open again in 1991. In 1992, outside serving was still banned in Manchester and night-time café licences were regularly refused because the Acting Head of Environmental Health thought that they would introduce the 'wrong sort of person' into the city centre. By 1993, Lovatt (1996) found that there had been a shift in regulatory attitudes in the city because Manchester's bid for the Olympics had highlighted the importance of culture and entertainment to the city. In 1994, admissions to both of Manchester's universities had risen by 30 per cent and 40 per cent of New Yorkers voted Manchester as the place that they most wanted to visit in a poll. The 24-hour city philosophy was given greater credence, and in 1993 Manchester magistrates and the City Council began to experiment with allowing pubs, clubs and bars to stay open later into the night and the early hours of the morning.

Between 1991 and 1994 club capacity in Manchester increased from 8,600 to 26,000. Twenty-five new café bars opened, together with eight new clubs and numerous restaurants (Lovatt 1996). A young student from south London identified a new opportunity for property development when he found that he was earning more from sub-letting space in a warehouse he had rented to store his posters, than from selling the posters at the street market.

Tom Bloxham went into business with an architect, an interior designer, a surveyor and two builders and founded a niche market in property through his new company, Urban Splash, regenerating industrial buildings into stylish loft apartments and bars in mixed-use development. Urban Splash's web-site proclaims: 'In the beginning there were factories. And they weren't working any more. But we thought they were beautiful' (Urban Splash 2008). Urban Splash are now occupied with regenerating developments across the UK, including an entire neighbourhood in Manchester.

Manchester was not the only city that experienced a substantial growth in its night-time economy. Leeds followed a similar trajectory (Lovatt 1994). At first it was the new entrepreneurs who were the 'urban pioneers'. As it became clearer that this was a new profitable sector, the structure of the alcohol and entertainment industry changed. Stock market deregulation in 1986, followed by a surge in the British economy following the decline of the late 1980s and early 1990s found new investors from Germany, the USA and Japan putting their money into the new pubcos (Chatterton and Hollands 2003). Brewery companies diversified into clubs and hotels.

These changes presented four different orders of entrepreneur. A former door security man, a 'bouncer', started what is currently Britain's largest nightclub company in 1987, Luminar plc. Luminar was listed on the London Stock Exchange in 1996 with a market capitalization of £30 million. In 2005, the company operated five brands, four of which were late-night, in 283 venues across the UK (February 2005 figures). Another entrepreneur with a slightly more exalted background, Jamie Palumbo, founded the Ministry of Sound, a famous club founded in 1991 in south London. As the son of the property developer Lord Palumbo, Jamie had some business advantages. The Ministry of Sound prospered and has become a 'brand' with its own 'philosophy' and provides fans access to its own distinctive styles of dance music via the media, radio, internet downloads and its own production with tours and albums, in addition to running venues in Egypt and Singapore as well as the original club at the Elephant and Castle in south London (Ministry of Sound 2008).

As nightlife became established in specific parts of certain town and city centres, a dynamic was set up as more companies and entrepreneurs saw their opportunities. Ultimate Leisure, for example, grew from owning a small number of run-down bars into a stock market-listed public company with a turnover of £30 million in the years between 1997 and 2001. The expansion of the night-time economy even bucked the trend for out-of-town development. In Sheffield, for example, the entertainment corporate giant, Rank, sold its out-of-town club and redeveloped a disused Odeon cinema in the city centre, conveniently situated between the taxi rank and the 'main drinking drag' (Hadfield 2006).

Changes to the structure of the industry did not benefit all of the cultural entrepreneurs who were the innovators of the 1990s. The re-structuring of

3.2
Digital Heaven.
Source: Toby
Price, www.
photosinthedark.
com, 2008.

the industry has assured the dominance of the major corporates. The reasons for this are easy to speculate about, although there has been little systematic work to date from the property point of view. As 'hotspots' emerged, institutional landlords realized that they could make a greater return on their leases with alcohol- and food-related uses, leading to rent rises. Investment has to be larger and becomes beyond the means of an individual or a group of friends. As the stock market becomes involved, each company or part of a company has to grow and show an increase in profits. This demands either a higher volume of sales, increased prices or reduced running costs. The logic is of ruthless competition, driving down prices and trying to gain a larger market share.

The expansion of the night-time economy in England and Wales was facilitated by two other factors. The start of this chapter mentioned the pressures on city centres. Planning authorities were keen, if not desperate, to make sure that their city centre was full and that there were no empty buildings to create an atmosphere of dereliction and decline. Although the Beer Orders placed restrictions on the numbers of pubs that a brewery could own, it did not place any restrictions on their size. As has been recounted earlier in this chapter, major stakeholders such as banks and insurance companies withdrew from the town centre to 'back office' sites, leaving architecturally handsome buildings with large floor areas. An independent developer and bar owner in Leeds described the movement of the major pub chains and brewers into these types of premises in Leeds city centre as a 'stampede' (Connolly 2005).

At first planning authorities welcomed an influx of entertainment uses, reasoning that these would not only bring jobs, but that they would also create a livelier town centre at night. Further on into the 1990s and into the next century, when faced with an over-concentration of licensed premises, planning authorities found themselves in greater difficulties because of the looseness in the planning regulations that permitted a small restaurant to be transformed into a café bar without the need for planning permission. Wily landlords and corporate leaseholders could then make this conversion and then apply for an expansion in size or for later hours. A café bar that provided a dance floor and could operate after the terminal hour of 11 p.m. in London, and 10.30 p.m. outside London, had an advantage over a nightclub because it did not charge an entry fee, nor was it subject to the attention of the police vice squad.

A further boost to expansion was given by changes in the rules that magistrates adhered to in the granting of liquor licences. In 1999 the Justice Clerk's Society published a Good Practice Guide that removed the concept of 'need'. This concept was one that magistrates applied when considering an application for a new licence. 'Need' applied to a judgement about the number of licensed premises that were appropriate for any given neighbourhood. The removal of a consideration of 'need' allowed potentially unlimited expansion (Hadfield 2006; Light 2005). Expansion is limited, of course, by the market itself. Although overall turnover figures seem impressive, comments from the operators themselves suggest that profit margins are tight. One of the two major insurers for nightclubs commented:

> Yes, in my personal experience what I've seen is the national oper-
> ators, such as the [name of major chain], and [two other major
> chains], have got a large amount of capital investment, from wher-
> ever it is they have got it from, and looked to gain a market share
> across the country, and the only way they managed to get that
> market share is by undercutting the existing local operators. They've
> cut each other's throats.
>
> (Insurer, nightclubs, interview July 2005)

It was not only in the nightclub sector that there was stiff competition. For example, a police report noted that Blackpool in 2004 had 60,000 'vertical drinking spaces and approximately 40,000 potential customers' (Lancashire Constabulary, Western Division 2004).

Regrettably, there are few accurate statistics of growth throughout the UK. Prior to 2005, statistics about licensed premises were collected nationally and by region and could only provide a consolidated picture. The statistics covered the number of premises and not their capacities. Before 2005, local authorities were not required to keep readily accessible records of their licensed premises and many kept paper records, using addresses as an index. This posed a problem for a spatial analysis. The dramatic increases in 'drinking spaces' per potential customer in certain city centre locations were not revealed. Similarly, the expansion of late-night activities was not recorded spatially and local planning authorities, such as the Greater London Authority had to commission a specialist report by Urbed and CASA (2002) to gain a better understanding of what was going on in their area.

Night-time 'clone towns'

The UK government had intervened in the alcohol industry to stimulate competition. Researchers at the University of Newcastle-upon-Tyne carried out an in-depth study of three cities, Newcastle, Bristol and Leeds between 2000 and 2002. They investigated the connections between an influx of youth into these 'cool cities' at weekends and the roles and attitudes of 'providers', that is, the venue owners, managers and the companies behind them and 'regulators', that is, city council officials, the police and other agencies (Chatterton et al. 2002). Their findings were that there had been a reduction of independent venues and small operators.

The structure of the industry that they reported was complex. Traditional pubs and 'ale houses' were joined by disco bars, 'themed' pubs or bars, 'style' bars, café bars and alternative pubs and bars. Behind this apparent profusion lay a relatively small number of operators, some of whom were national, if not trans-national, others who operated across the local region and yet others who were independent. There was a dramatic increase in the numbers of 'managed' premises, in all venue types. This means that the parent company effectively has control over the venue, setting targets and taking decisions as to décor and pricing. The increase was at the expense of 'tenanted' premises, where the pub or bar is also the publican's home.

The report noted the increase in 'brands' in each city. The brands tended not to be that of the pubcos or the breweries, but were related to the style and type of venue. These are aimed at a particular niche market, which might be sports fans, students or fans of a particular style of music. These brands appear in particular towns and cities across the country. In the snapshot that the research team presented of their three cities, Leeds had approximately 30 per cent of independent venues in its city centre, whereas Newcastle-upon-Tyne had only 5 per cent.

The market in the alcohol-related entertainment industry is volatile. Transfer of brands between different ownerships is commonplace. A sense of the rapid rises and falls in the market can be seen in the way in which Luminar, whose annual report claims it to be the UK's leader in late-night entertainment, consolidated its operation to 50 venues in 2007, with a strategy to increase this to 120 by 2009 (Luminar plc 2007). A snapshot of the ownership of brands commonly found in UK high streets is set out in Table 3.1. This table includes

Table 3.1 **Corporate ownership of nightlife in the UK February 2008**

	Category	Brands
JD Wetherspoon	Pubs (687) and hotels	Wetherspoons, Lloyds No. 1
Mitchells & Butler	Pubs and bars	Edwards, Ember Inns, Flares, Goose, Hollywood Bowl, Nicholson's, O'Neills, Scream, Sizzling Pub Co.
Enterprise Inns	Pubs leased and tenanted (7,000)	Individual pubs, urban and rural
	Restaurants	Alex, All Bar One, Browns, Express by Holiday Inn, Harvester, Innkeepers' Lodge, Toby Carvery, Vintage Inns
	'Classics'	Pubs with long architectural and trading history
Punch Taverns (Spirit Group)	Managed pubs (800)	Individual or groups of pubs
Regent Inns	Australian sports bars (50)	Walkabout
	Comedy clubs (16) and franchised brand	Jongleurs
	Youth bars (9)	Bar Risa (linked to Jongleurs)
	American-style restaurants (31)	Old Orleans (29); Quincy's (2)
The Restaurant Group (also TRG concessions in shopping centres and airports has a wider list of brands)	Restaurants, pub restaurants, café bars, pubs (300)	Chiquito, Frankie and Benny's, Garfunkels, Bluebeckers, Edwinns Brasserie, Brunning and Price
Pizza Express	Pizza restaurants (300)	
Luminar plc	Nightclubs	Oceana, Liquid, Lava & Ignite, Life, The Jamhouse
Premium Bars & Restaurants (formerly Ultimate Leisure)	Bars, restaurants and clubs (50+) and two hotels	The Living Room, Prohibition, Bel and the Dragon, others
Inventive Leisure	Vodka bars (53)	Revolution
Tragus Ltd	Restaurants (French, Italian, Spanish) (240)	Café Rouge, Bella Italia, Strada, Ortega, Brasseries
Whitbread	Restaurants (433) and hotels	Brewer's Fayre, Beefeater
Novus Leisure (formerly Urbium)	Nightclubs (9) and bar-restaurants	Tiger Tiger, others
Glendola Leisure	Bars, clubs and restaurants	Waxy O'Connors, The Loft, The Terraces, Rainforest Café, others
Yum Restaurants International	Restaurants	Pizza Hut (600), KFCs

Source: company web-sites (accessed 15 February 2008).

restaurants but does not include McDonald's, many of whose franchises operate late at night. The table illustrates how the typical night-time high street may appear to offer a multiplicity of independent venues, but is in fact controlled by a much smaller number of companies. The larger corporate owners are registered on the London Stock Exchange, the smaller on the Alternative Investment Market (AIM) and others, such as Glendola Leisure Ltd, are held in private ownership. This leads to different investment strategies and impacts. In addition, the size of the venues owned by some of the smaller companies belies their apparent lack of influence. For example, clubs under the Tiger Tiger brand, owned by Novus Leisure, have a minimum capacity of 1,000. The Glendola Leisure Ltd website asserts that their major bar chain, Waxy O'Connor's, was the first 'superpub', that is, a very large pub, to be opened in the UK.

This structure of ownership is particular to UK town centres and provides a mirror image of the day-time. The appearance of 'clone towns', dominated by retail chains, has been the subject of independent and parliamentary reports (Simms *et al.* 2005). That the night-time town centre should be similar is not, in itself, surprising. It does seem, though, that this pattern of ownership is not generally repeated in the 'informal' nightlife sectors in mainland Europe and the USA. In Denmark, drinks providers are prohibited from having shares in venues and therefore ownership is mainly independent. In the USA, although major corporations dominate the 'planned' 'urban entertainment districts' in the downtowns of major cities, where a spontaneous informal nightlife street or quarter has developed, it tends to be in multiple ownerships (Campo and Ryan 2008).

The urban environmental impacts of the competition between different brands will be considered in greater depth in the next chapter. Suffice to state, there is no evidence in Britain that it has led to a relaxed style of consumption. Rather, in the period 1995–2003, market competition assisted in producing a 'demographic ghetto' in certain micro-districts in many British towns and cities, where loud, noisy venues full of 16–30-year-olds standing up, dancing and drinking to excess became the norm. This was not the style of consumption envisaged by planners when they first embraced the notion of 24-hour cities, cultural quarters and 'creative' quarters. The image of the 'creative' neighbourhood did not include hen and stag parties snaking their way down the street, advertising placards declaiming the latest rock-bottom bargain prices for drinks offers, or the spectacle of waitresses in bikinis selling single 'shots' of spirits to the lengthy queues outside venues.

Concluding comments

British urban planners were confronted with the conundrum of how to revive declining town centres in the late 1980s. The ideas offered by the concepts of the 24-hour city, cultural quarters and creativity offered an urbane alternative that could bring Britain in line with its mainland European counterparts. The vision of lively neighbourhoods populated with gentrified residential units, small clubs and bars, good places to eat and a blurring set of activities spanning cultural production and consumption was compelling.

Changes in the structure of the drinks industry at first seemed to facilitate revitalization. For the major brewers who were restricted in the number of properties that they could own, taking over an empty banking hall offered economies of scale. Disused warehouses and industrial buildings initially provided cheap alternative premises for new entrepreneurs who could fill the space with the latest DJs and cutting-edge music. As the possibilities became more known, competition heightened. Investment companies looking for new markets saw their opportunities and moved in. The ownership of night-life became big business, with shareholders expecting dividends and good out-turns. Planning, in the absence of alternatives, simply followed demand. The impact of the expansion of a specific form of late-night entertainment proved to be both positive and negative. It is this that we shall turn to in the next chapter.

Chapter 4

Party cities

Most of all, I love Manchester. The crumbling warehouses, the railway arches, the cheap abundant drugs. That's what did it in the end. Not the money, not the music, not even the guns. That is my heroic flaw: my excess of civic pride.

> Quote from *24-Hour Party People*, attributed to Tony Wilson, Director of Factory Records, journalist, impresario and founder of the Hacienda Club

The West End at night is a pretty squalid place, just horrible.

> Representative, Soho Society, London 2002

The previous chapter explained the expansion of evening and late-night entertainment as a product of the 'cultural turn' in urban policy. It suggested that in the UK, growth in this sector has been driven to excess by the opportunism of entertainment providers and fragmented policy objectives. This chapter will expand upon this argument and explore the impacts of expansion on the experience of place, its positive attributes and its negative externalities. Much of the focus of the chapter will be on experiences contained within Britain and Ireland. The UK experience is not unique and other cities in the world have different experiences of expansion, shaped by their specific cultures, economic context and histories. Helsinki is an example of a northern European city that is adopting a planned approach to expansion and provides a comparison to what appears to be an unplanned and informal growth of entertainment areas in some cities in the USA.

There are many different ways in which to consider the impacts of expansion. For city councils, new economic activity is welcome. Not only does the appearance of a city in decline fade from memory, new enterprises bring jobs, revenue, income and people into the city centre. There are also the supply chains, boosted by an increase in activity. Local residents who live close to new venues or later-opening venues are likely to have a different point of view. Noise, anti-social behaviour, litter and crime become key issues. There is also the impact on residents living in the wider urban region and tourists who may either be attracted or repelled by changes to the image of night-time activity in urban centres. Other businesses operating in an area of expansion, a 'hotspot', may also have concerns about the impact of late-night revellers on their operations. Urban theorists, following a political economy framework, raise more abstract objections that can be classified according to different concerns, ranging from the exacerbation of social difference and social exclusion to the extinguishing of local identity (Latham 2003).

The weight of negative responses to the night-time city poses difficulties for commentary. On the one hand, as will be demonstrated, the negative consequences of an unregulated expansion of late-night entertainment uses cannot be dismissed as merely a 'suburban' response on the part of city centre residents or an analysis produced by middle-aged, middle-class academics removed from the excitement and spontaneity of youth culture. On the other hand, it is easy to fall into a position of negativity and fail to acknowledge the benefits of new forms of sociality and engagement. The discussion that is presented here is not intended as a kind of 'balance sheet' for the night-time economy, weighing up the positives of jobs created versus the negatives of expenditure on extra ambulance services. An expansion of nightlife offers new types of activity that, in conducive circumstances, can increase the potential for human exchange, understanding and pleasure. This potential is restricted and at times extinguished by other factors, however, such as the economic context for provision, the type of regulatory environment and a lack of forethought in spatial planning. Finally, there is the addictive nature of alcohol and other drugs, the cost of which must be fully acknowledged. This last issue will be considered in further depth in Chapter 5.

Liminal spaces

Theorists of postmodernity have commented on the carnivalesque nature of contemporary consumption, comparing the disorientation many feel in crowded shopping malls, hotels and department stores (Featherstone 1991; Jameson 1991; Watson and Gibbs 1995) with that of a street festival. This comparison is somewhat strained, especially when contrasted to the sensations of walking around a night-time economy hotspot in Britain on a summer evening. Here the sensation is of the street being a socially unbounded space, in suspension from its everyday realities of commercial transaction, offering possibilities and promise for leisure, play and entertainment. For some participants the possibilities include a sexual encounter, even leading to a life change. For others it offers simply a good time, a chance to get away from the responsibilities of the week, or to get 'out of their heads'.

Campo and Ryan (2008) draw on Foucault's notion of heterotopia, or 'other spaces' (Foucault 1986) to describe this state of being. They suggest that rather than being utopian or dystopian, specific sites exist as a temporal and physical space in which social inhibitions are loosened and social norms are reconfigured. In a study of the unplanned growth of nightlife in the marginal neighbourhood of Water Street in Milwaukee, USA, for example, Campo and Ryan argue that the type of middlebrow entertainment provided by independent cultural entrepreneurs to young city dwellers and suburbanites is heterotopic. They distinguish this type of area, with its small bars and taverns in run-down, old buildings, from the corporately styled 'urban entertainment districts' analysed by Hannigan (1998). Urban entertainment districts are characterized by large-scale buildings that turn inwards, providing sanitized accommodation for corporate chains. By contrast, Campo and Ryan designate Water Street as an 'entertainment zone'. Entertainment zones are found in as many as 30 cities in the USA, with each sharing common characteristics: small-scale buildings accessed in a pedestrian-friendly manner and a variety of architectural styles. They can be in mature neighbourhoods such as East Village in New York, or even found amongst second-order retail strips such as the Richmond Strip in Houston (Campo and Ryan 2008).

Campo and Ryan are enthusiastic about these unplanned entertainment zones and argue that they offer benefits to city governments as examples of 'everyday urbanism'. For some Americans, they suggest, this may be the only contact they have with ordinary urbanism, since the rest of their life is spent between business parks, retail malls and suburban sprawl. This argument for a more European style of urbanism has echoes of the promotion of cultural quarters and 24-hour cities in the UK.

As explored in the previous chapter, planners and politicians 'boosted' their city centres as locations for consumption, drawing on comparable mythologies of a bohemian past and a supposedly 'European' present. One resident of Soho in central London summed this up as a fantasy:

> Part of the problem here is this strange myth that Soho is where you can behave disgracefully and it's a kind of never-never land where bad behaviour is accepted. All these people are living in the suburbs, bored out of their minds, and maybe they've been to Ibiza, and they think, 'ah well, let's go and let it all hang out in Soho'. And of course in so many of these joints various kinds of drugs are available, and that's known all over the UK and all over the home-counties and in Europe. So, it's a dangerous 'sin' place to come and hang out with all your peers.
>
> (Editor, *Soho Clarion* 2002)

The effect of drink and drugs on the individual is – of course – intoxication, altered states of awareness, heightened perceptions of pleasure and, ultimately, when consumption has been taken to excess, disorientation and worse. With regard to experience on the streets, the impact of so many people in different states of intoxication, combined with the sounds of music coming from the discos and bars, is to contrive an impression of a party happening outside.

Whether we refer to these sites as party cities, entertainment zones or heterotopias, these late-night destinations offer a spur to economic growth and regeneration. For the venues, a 'party' atmosphere on the streets will encourage customers to loosen inhibitions, enjoy themselves and above all, spend. For many city councils, the transformation of once empty streets from places of desolation to scenes of memorable enjoyment is equally welcome. The promotion of party cities has since become a recognized specialization of the tourism and leisure industries. The discussion that follows develops our previous discussion of the cultural turn in light of the growth of the party city. It examines various examples of cities that are pursuing this path as a means to re-imagine their city centre, with both negative and positive effects.

Economic benefits of the night-time economy

The principle benefit of the party-city ethos is monetary. Different associations of licensed operators are eager to highlight the extent to which their activities support central government revenues. The British Beer and Pub Association, for example, estimate that in 2006 the total taxation on beer, including VAT, contributed £5,903 million to government coffers (British Beer and Pub Association 2008). The Portman Group, which represents drink producers and is concerned with social responsibility, cite the government's own Alcohol Harm Reduction Strategy in estimating that the alcoholic drinks market produces over £30 billion in wealth for the UK economy (Portman Group 2008). The total turnover for the night-time economy is about 3 per cent of GDP (Hobbs 2005a).

A further benefit of the expansion in the evening and late-night economies is the expansion of jobs. Research conducted by the authors for a UK charity group, the Civic Trust, found that 37 per cent of local authorities surveyed in a national sample from England and Wales reported that they agreed or strongly agreed with the statement that expansion in the evening and night-time economy in their area had brought economic benefits in the form of an increased number of jobs (Roberts and Gornostaeva 2007). According to the Labour Force Survey 2005–2006, over one million people are now employed in hotels, pubs, bars, nightclubs and restaurants in the UK (HM Government 2007:

34). There was a 19 per cent growth in jobs in pubs, bars and nightclubs in Britain as a whole between 1995 and 2001 (GLA Economics 2003: 43).

London's share of the leisure economy is greater than its share of the economy as a whole. Increases in jobs had been especially marked in certain growth areas in Greater London. In a comprehensive report, *Spending Time: London's Leisure Economy*, the Economics Division from the Greater London Authority marshalled an impressive body of evidence to demonstrate that increases in the numbers of restaurants, cafés and fast-food takeaways had resulted in 122,000 outlets operating in 2002, bringing in a turnover of £4.7 billion annually. Of these, 12,000 were restaurants, and in the period 1995–2002, their numbers had increased by 28 per cent. In the pub, bar and nightclub sector, jobs had increased by 37 per cent in the period 1995–2002. Although there were 58,000 venues in these categories in Greater London in 2002, they made up only 11 per cent of Britain's employment in this sector and were less important to London's local economy than they are nationally. While eating and drinking made up three-quarters of London's leisure economy, entertainment, gambling and visitor attractions are significant. Theatres, cinemas and live performance venues accounted for 44,000 jobs in 2002 (GLA Economics 2003).

Increases in the numbers of evening and night-time economy jobs in other major city centres were less dramatic, but significant. One estimate was that 12,000 extra jobs were created in Manchester for bar, waiting and security staff (House of Commons 2003b: Ev 16). The capacity of licensed premises went up by 242 per cent in the period 1997–1999 (Home Office 2007b). Town Centres Ltd estimate that a licensed restaurant of up to 560 m^2 typically employs up to five full-time and 25 part-time staff, and a nightclub (1,860–2,800 m^2) between 50 and 90 staff, of whom half would be full-time and half part-time (Town Centres Ltd 2001: 29). These figures provide some multipliers for assessing the number of jobs that have been created in British towns and cities. While precise figures are not readily available, the graph below demonstrates that in the period 1994–2004, the number of licensed premises in Leeds city centre doubled. At the time of writing, Leeds city centre accommodated 308 licensed premises with licences allowing them to trade after midnight. Figure 4.1 provides further detail of licensed premises in Leeds (Goodall and Lawrence 2007). In 1997, the licensed capacity of Nottingham's small city centre was 61,000; by 2004 that had risen to 108,000, while Manchester city centre had a capacity of 55,000 in 2003 (House of Commons 2003b: Ev 20), with 700 licensed clubs and bars. These figures should be assessed against a 30 per cent national rise in the number of licensed premises from 1980–2005 (Hobbs 2005a).

The types of jobs created by an expansion in the food and drink sector are varied, but outside of London the majority are part-time (GLA Economics 2003). Many are also transient and low-paid. Nevertheless, the skill

4.1
**Licensed
premises in Leeds
city centre.**
Source: Goodall, T.
and Lawrance, K.
(2007: 20).

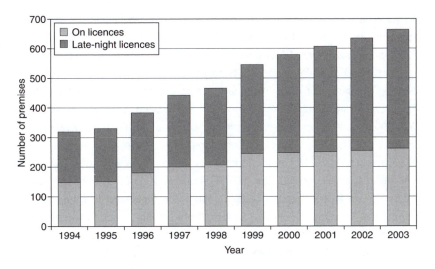

4.1
**Licensed
premises in Leeds
city centre.**
Source: Goodall, T.
and Lawrance, K.
(2007: 20).

4.2
**Overlap between
different
'economies'.**
Source: adapted
from Evans 2001:
41 by authors.

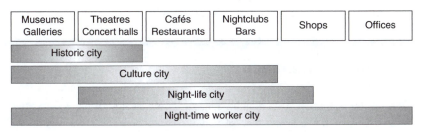

base is rising as organizations such as the British Institute of Innkeepers introduce new sets of qualifications for responsible bar-tenders and the Security
Industries Authority requires training and accreditation for door staff. Some
innovative regeneration agencies have also introduced training schemes for
high-class restaurant staff of which the Hoxton Apprentice, run by the social
enterprise charity Training for Life and the Shoreditch Trust, is a prime example.
In addition to those directly working in night-time economy-related businesses,
other economic activities are stimulated, such as the fashion and music industries, advertising, merchandising and public relations. More demands are also
placed on other services, such as hotels and transport. As a report to the Arts
Council England argues, the locally produced live performances are a key driver
of the night-time economy, attracting tourism and visitors. The synergies
between the various aspects of the local economy are set out in Figure 4.2.

Spatial intensities

The rapid expansion of the night-time economy in certain town and city centres
has drawn unprecedented numbers of people into these places. Greater

Manchester Police estimate that there are 120,000 people on the streets at night on weekends in Manchester city centre (Greenacres and Brown n.d). The north-eastern city of Newcastle-upon-Tyne accommodates 80,000 people on Friday and Saturday nights in its town centre. The chief police officer in Nottinghamshire told a parliamentary select committee in 2004 that on a Friday night in December, there would have been between 80,000 and 120,000 people out on the streets in the centre of Nottingham, with only 40 officers to police them (House of Commons 2004). A similar number, between 80,000 and 120,000, frequent Leeds city centre on weekend nights. A smaller centre, Kingston-upon-Thames in the suburbs of London, accommodates 12,000, also at weekends. These numbers are replicated in other centres across the UK. One survey of the total visitor numbers added by the night-time economy to different sizes of settlement noted a variation between 80,000 and 150,000 (Social Issues Research Centre 2002: 37). The extent of this variation demonstrates not only the scale of expansion, but also its variation across six regions of the UK.

Youthful dominance

Despite the economic and social benefits of the expansion of night-time activities in Britain, the statistic that all surveys of consumers in the night-time economy are agreed upon is the domination of the high street by the youth market (Mintel 2004, 2006a). This is not a necessary outcome of the night-time economy, but the situation in Britain today is that young people (aged 18–24) are the group most likely to visit high street venues. A GfK NOP survey found that over three-quarters of this age group had visited a high street chain or themed bar, pub or club in the previous 12 months. The proportion of the population going out declines with age, and figures supplied by the UK government estimate that while 45 per cent of 16–35-year-olds have a night out once a week, only 15 per cent of over 55s do (ODPM 2005c). The concentration of youth led to market researchers in a 2004 report referring to the 'demographic ghetto' of the high street (Mintel 2004). While more recent market research suggests that some of the major chains are now looking to attract an older and female market in order to retain their base, the observation of youthful domination remains (Mintel 2006a).

These findings have been supported in studies conducted in the south-west of England and Wales, in Cardiff, Swansea and Bristol. This research has made significant findings with regard to the patterns of use for different types of venues within the evening and night-time economies and the spatial patterns that have emerged at different times during the evening and night-time. The observation that older people are deterred from visiting British city centres at night was confirmed by a series of visitor and household surveys carried out in Swansea and Cardiff in the late 1990s (Bromley *et al.* 2001; Thomas and Bromley 2000: 1420). Almost two-thirds of those interviewed did not feel safe in the city centre alone at night, although over 80 per cent would feel safe when accompanied. In Swansea, where the city centre offered at that time little beyond retail and entertainment uses, 52 per cent would avoid shopping after 4 p.m. in the afternoon, whereas in Cardiff, where the university and many offices are located in the city centre, 29 per cent were also deterred (Thomas and Bromley 2000: 1420). A study of Swansea found that 73 per cent of the under-30s in the study sample visited the city centre at night once a week or more, in comparison to 28 per cent of the 30–50 age group and 11 per cent of those over 50 (Bromley *et al.* 2003: 1838). Both city centres tended to have a greater preponderance of middle-aged affluent people in the early evening, but in the later hours before and after midnight, younger, less affluent people dominated the streets. Older people tended to avoid these 'drinking streets' because of a fear of crime in the late evening.

A later study of Swansea city centre probed more intensively into precise spatial configurations and their occupation over time (Bromley *et al.* 2003). Swansea has two streets where late-night bars and nightclubs are concentrated. The major cultural attractions of theatres and cinemas were not located adjacent to the hotspots. During the day-time, because of the lack of retail activity, these 'drinking streets' were less well used. The evening and night-time patterns of use over time revealed a complex web of activity. Late-night shopping occupied the early part of the evening, from 5.30 p.m. until 9 p.m., then restaurant, café, theatre and cinema customers drew in people from 5.30 p.m. to 11 p.m., pubs and bars from 9 p.m. to 2 a.m. and nightclubs from 11 p.m. until beyond 2 a.m.

Within the groups who would venture into city centres either frequently or occasionally, visitor, resident and household surveys revealed age and income differentiation in terms of their chosen night-time activities. Surveys of residents and on-street residents and visitors during four cultural events in one year in Swansea found that the most popular venues were pubs, bars, restaurants and clubs with some cinema attendance. Desire to attend 'high' cultural institutions was in a minority (Tallon *et al.* 2006). Surveys of city centre residents in Bristol and Swansea found that it was those residents in the top two upper income groups who were least likely to visit nightclubs, but were more likely to visit restaurants, while for the next to lowest income class (fourth out of a gradation of five), the reverse was true. Interestingly, the same proportion of all income groups in the city centre, nearly half, visited pubs once or more times per week. This is not to say, as the researchers point out, that the pubs themselves are not differentiated by status and type (Bromley *et al.* 2005). Nationally based market research confirms the finding that city centre bars and pubs attract customers from all but the lowest income groups, whereas theatres attract older, more affluent audiences.

These findings led the researchers to conclude that the city centre had different space–time layers of use, leading to speculation about whether a mixing of late-night youth venues and venues that attracted older people, such as restaurants and theatres, was appropriate or desirable. The concern that late-night youth culture might be deterring older people from going out in the evening is not confined to the south-west of England. Our survey of local authorities found council officers making comments such as: 'A smart restaurant near the core would fail unless their customers are deposited and picked up from the door of the establishment.' Another from a spa town commented that 'tourists do not venture out at night' (Roberts and Gornostaeva 2007). Not only council officers voiced these misgivings. In an interview with a left-wing magazine, a junior government minister for the Home Office commented:

> I now find myself going to early evening cinemas and going to matinee performances at the theatre because I don't want to be in Manchester city centre on a Saturday night at nine o'clock. This means that the city is not attracting people like me with money to spend.
>
> (Anonymous 2004b: xi)

Violent crime

As is discussed in Chapter 5, violence and other offences occur as a result of a confluence of contributory facts, and alcohol alone cannot be held solely accountable. However, evidence points to a strong relationship between a concentration of licensed premises and anti-social disorder. In what later would prove to be a controversial report (see Chapter 6), Marsh and Kibby (1992) set out a 'recipe' for the production of violent crime and disorder associated with drinking:

> Firstly, we would encourage young men to drink large volumes of beer or other alcohol in a short period and in a traditional, macho style where such patterns of consumption and 'manliness' are reinforced by the marketing and advertising of the products. At a fixed time, well before most of the participants are motivated to return home, we would close the places that sell drink. Before doing so, however, we would encourage a peak of consumption, resulting in a peak of intoxication just prior to leaving. At closing time, we would roughly expel the drinkers onto the street, suddenly increasing the density of people and maximising the potential for friction and conflict. We would ensure that there was very little in the way of trans-

portation so that people would have to queue for the few buses and taxis and remain in the town centres even longer.

(1992: 20–1)

These conditions were apparent in many town and city centres as their night-time economies expanded. Although the 'recipe' appears exhaustive, other 'features' might be added to the mix that exacerbated the potential for aggression, scuffles and outbreaks of violence. These were: queuing for fast-food (Budd 2003), lack of police officers to deal with any incidents as they 'kicked off' (ACPO 2005) and the presence of litter and a general perception of lawlessness (Hobbs *et al.* 2001). Added to this, unlicensed minicabs touted for business and preyed on lone women. The scale of the problems at the turn of the millennium were revealed by headline statistics. For example, around 13,000 violent incidents were taking place in or near licensed premises each week (Deehan 1999), and in 2000 19 per cent of all violent incidents took place in or around pubs and clubs, including 33 per cent of assaults by strangers and 22 per cent of assaults by acquaintances (Kershaw *et al.* 2000).

Chapter 6 provides further analysis of the relationships between crime, disorder and urban centres, and Chapter 8 of the responses made by local and central government and other agencies. The point to be made here is that as the night-time economy expanded, alcohol-related crime and disorder increased and it took some time before the organs of governance 'caught up' and took measures to reduce it. This 'catch-up' operation is ongoing at the time of writing, as is evidenced by a representative of the Police Federation telling a House of Commons Select Committee that some small towns are like the 'Wild West' in the early hours of the morning (House of Commons 2008).

In 1989, 42,900 people were arrested in England and Wales for drunkenness. This fell to approximately 21,100 people in 2004, with a further 13,500 cautioned. The Home Office's own report into alcohol misuse, *Safe. Sensible. Social.* notes that since 1995 violent crime in England and Wales has fallen by 43 per cent. Though this reduction is not reflected in public perception, it translates into a wider reduction in alcohol-related crime. Again, according to *Safe. Sensible. Social.*, 'the actual number of offences where the offender is believed to be under the influence of alcohol has dropped by a third since 1995' (HM Government 2007: 22).

Nonetheless, crime remains an important deterrent for older residents to visit their town or city centre, and has become an overriding narrative of the late-night hotspot. A much-cited figure from 2004/2005 states that 48 per cent of violent offenders were thought by their victim to be 'in drink'. This statistic, from the British Crime Survey in England and Wales (2004/2005), has been elaborated by more detailed studies that examine the links between criminal behaviour and alcohol misuse – most of which have worked their way into the popular press (Social Issues Research Centre 2002; Hughes *et al.* 2007; Richardson and Budd 2003; Parker and Williams 2003).

The links between violence and alcohol consumption are complex, but the over-concentration of venues and the preponderance of intoxicated people in town centres at night, combined with strained infrastructure and services, makes for a potent mix. For our purposes here, the rapid expansion of the party-town ethos has also come to represent a decidedly contrary narrative to the equally powerful notion of the urban renaissance. That is, the desire to create vibrant, inclusive and creative 24-hour cities, as discussed in Chapter 3, has been challenged by the resulting situation of a commercially driven, alcohol-centred and youth-dominated city centre.

Tensions in city centre living

The previous chapter pointed out that although 'the urban renaissance' had encouraged a return to living in the city centre, the stereotypes of the new dweller as being a young, affluent party-goer was not entirely accurate. Further, a few UK towns and cities have maintained a substantial city centre population throughout the twentieth century. Cities such as London, Edinburgh and York have retained their city centre residential quarters, with some occupying peak levels of desirability, such as Mayfair and Belgravia in London's West End and the New Town in Edinburgh. Smaller cities, such as Brighton, also have attractive housing within their centres.

This observation is important to the debate about the impacts of an expanded night-time economy on city centres. Superficially, the twin policy objectives to increase city centre living and to expand the night-time economy might seem compatible, if city centre dwellers are young, affluent and consumerist. The ideal, as one Manchester city councillor put it, of the 'Mars bar' city, where work, rest and play take place within a compact city core represents a late twentieth-century update of EU strategy. Even within this concept, however, a closer examination reveals tensions. Young professionals, students and mature middle-aged professionals may enjoy 'going out', but as the previous discussion of the space–time layers of the city demonstrated, each group has different entertainment and cultural preferences, some of which are incompatible. A couple enjoying a quiet meal at a restaurant and a stroll home at midnight has an altogether different impact to a boisterous group of friends out on an alcohol-soaked circuit of the city's late bars. The impact of this last group is also distinct from that of a couple or a small group of music aficionados who spend a night dancing and taking illegal drugs at an independently provided themed club. The presence of the second is likely to have an adverse impact on both the restaurant-goers and the dancers in terms of noise, low-level disorder, actual violence, litter and threat.

Furthermore, the pattern and intensity of activities changes by the day of the week. An investigation by the police and the local authority in New-castle-under-Lyme, a market town in the West Midlands, found that different activities 'peaked' on different nights of the week in the period 2002–2003. Each activity was associated with different categories of disorder. Wednesday was student night – offences tended to be concentrated around the consequences of rowdy behaviour, such as vandalism, noise and street fouling. Friday was couples night – here the tendency was for fights to break out if one partner in the couple accused a partner in another couple of either 'ogling' or insulting their partner. Saturday was a night for groups of young singles, especially hen or stag parties, and was associated with violent affray between large groups.

The evidence from different surveys amongst city dwellers also suggests that many do not participate in the night-time economy within their areas. A survey of residents in Bristol, Cardiff and Swansea found that 30–40 per cent of respondents in those cities either 'never' participated, or did so 'not often' (less than once a month) (Bromley *et al.* 2005: 2416). A survey of participants in the night-time economy in Shoreditch, Hackney, in east London, found that only 21 per cent of the sample lived within a ten-minute walk (Urban Practitioners 2004: 28).

Expansion in the night-time economy has also caused tensions with residents due to noise, litter and anti-social behaviour. A House of Commons inquiry heard evidence from residents in Bath, London's West End, Headingley in Leeds, Manchester, Camden in London, Nottingham and Bristol of their concerns and frustration about these issues (House of Commons 2003b). The evidence given by the Federation of Bath Residents' Associations described these problems most eloquently:

> Right in the heart of the historic centre there are 45 licensed restaurants, 49 bars closing at 11pm, and 12 bars with late licences serving alcohol until 2am, six nights a week.... All 106 are mixed in with hotels, hospitals and homes, and obtained their planning permissions long ago in quieter days.
>
> The combined capacities of the 2.00am premises total 3,742 patrons and they are all situated in an area measuring 800 metres × 400 metres.... In this zone there are an estimated 2,000 people sleeping, half of whom are permanent residents. A Council survey found that 77 per cent of residents are affected by the street noise of nightlife (laughing, shouting, screaming, crashes, cars starting up etc).... In Abbey Ward, which contains these nightspots, disorder and violence are seven times and damage is 40 times higher than the average district level of crime.
>
> On busy weekend nights (Thursday–Saturday) boisterousness and noise commences mid-evening as intoxicated youngsters go

> circuit drinking getting tanked up on cheaper booze before going on to the clubs. Noise continues for hours until the city streets are finally empty – around 3am.... In the mornings there are not enough staff, due to 'lack of funds', to clean the city streets to achieve even the minimum standard as set out in the Environment Protection Act. Broken glass, litter, vomit and urine are often to be seen.
>
> (House of Commons 2003b: 33)

This description sets out the typical complaints that residents make. There is, however, a level of resistance to 'hearing' this on the part of commentators. Hobbs comments that there was a reluctance on the part of the authorities, particularly the police, to take problems associated with the night-time economy seriously prior to the millennium (Hobbs *et al.* 2003). For planners and city managers, the evidence disturbs the city boosterist narrative of the urban renaissance. It also acts to the detriment of city marketing and the goal of attracting visitors into city centres. In addition, there is also a punitive attitude towards city dwellers that draws on a rhetoric of choice. For example, a licensee interviewed in a study of hotspots remarked:

> I think people who live in town centres are going to have to put up with that. I don't think they should be saying 'oh, you're too noisy, you're going to have to close your doors'. At the end of the day they bought flats in an economy that's already going. So it's tough luck.
>
> (London borough licensee, interview 2006)

In a more reflective manner, the architectural critic Rowan Moore commented that late-night opening in London's West End was to be encouraged, otherwise London would have a 'Cinderella' quality of closing before midnight (Moore 2000).

The reluctance to recognize and give weight to residents' complaints about disturbance and noise fails to take account of the way in which the street environment can move from a compatible level of urbanity and liveliness to one which is threatening and incompatible with normal residential life. This evolution happens as the party city and the urban renaissance come into contact. A prime example of this is provided by the changes that took place in Old Compton Street in Soho, in London's West End, between 1991 and 2003.

Old Compton Street

In Soho, the dilemma of living the urban renaissance was felt particularly acutely. The first 24-hour city conference heard police evidence in 1993 that

Soho was the first neighbourhood to successfully operate as a 24-hour area. Westminster City Council adopted a relaxed attitude to public entertainment licensing and this allowed venues to open until 3 a.m. and 4 a.m. In 1999, 149 premises held music and dance (late-night) licences, an increase of almost 70 per cent since 1995. This almost quadrupled capacity from 33,418 to 127,860 (Town Centres Ltd 2001). In 2003, there were 263 late-night licences in Soho. Almost 9,000 homes had been built in Westminster City Council's area of jurisdiction since 1991, of which over 2,500 were in the city centre. One resident estimated that there were 5,000 people living in Soho at the millennium.

Residents opposed the granting of new planning permissions for licensed premises, the issue of new licences and changes to the hours and conditions of existing ones. On occasion, they even took direct action, as this longstanding Soho resident attests:

> One of my earliest experiences was when I was living in Archer Street. I was living next door to Paramount City, bought by Paul Raymond. He let Paramount City out to this character who put his son in charge of running a disco – it was to be the first country and western disco in the West End. They put a lot of money into it, they did line dancing classes – it wasn't a problem at all. Of course it was a financial failure and they quickly reverted to other kinds of music and they turned the volume up. Where I slept, the other side of the wall there was this disco. And I got into a real war with those guys – I called the noise team out every night. One night, I was woken up at 3, and I put a coat on over my Pyjamas, I grabbed my hot water bottle – it was something like November or December, and I went around there, unplugged my hot water bottle and emptied the contents all over the reception desk. A couple of heavies came up to me and then the manager appeared, looking rather flustered in the background and called them off. There are times when you can only make a statement to somebody, it's no good being rational and writing letters because you have to let people know how you feel.
>
> (Editor, *Soho Clarion* 2002)

By 1997, residents felt that the noise and disruption had reached unacceptable levels and applied to the High Court to take the Council to judicial review. Following this, the Council rapidly reviewed its policies and declared that Soho should be part of a West End stress area, in which no new permissions should be given or extended hours for liquor licences granted for restaurants, clubs or bars (Westminster City Council 2001).

An investigation conducted by the authors examined the impact of Soho's late-night culture on its street environment (Roberts and Turner 2005).

4.3
Soho at 8 p.m. The crowds and traffic continue into the early hours of the morning.
Source: captured from video taken by authors.

Old Compton Street is in the heart of Soho, whose reputation as Britain's most famous red-light district goes back to the nineteenth century, becoming known as the 'sleazy square mile' after the Second World War. By the 1950s, Soho had acquired a reputation as a bohemian quarter. In the 1980s, when the area became gentrified with the influx of media companies, it was marketed as having a 'continental' air and there was a trend for a continental café culture and alfresco drinking and dining. Mort (1995) considered the external appearance of tables and chairs to be 'a distinctly English reading of Gallic culture'. The term 'English reading' is particularly apt in the context of this book. In 2002, over 60 premises in Soho held night-café licences, which permit opening beyond 11 p.m., and one of the cafés on Old Compton Street stayed open until 5 a.m.

In the words of a local resident, in the 1990s Old Compton Street was transformed from a street of delicatessens into a street of bars. This perception was supported by a study of the growth of entertainment uses in the West End. Between 1992 and 2002 there has been a 35 per cent growth in food and drink (A3) uses, a loss of 350 retail units, a loss of 7 per cent of office floor space and a 35 per cent growth in late-night capacity of entertainment uses. There was also a 40 per cent growth in liquor licences in the South Westminster area (Town Centres Ltd 2001). At the time of the research conducted by the authors, there were six clubs and bars with late licences and a total capacity of 1,773 customer places in Old Compton Street alone. In addition, there were pubs, two delicatessens, a theatre, restaurants, late-night convenience stores, shops, offices and flats.

Although crime and anti-social behaviour are and were undoubtedly problems in hotspots such as Soho, it would be misleading to blame all the problems on a small number of violent or criminal individuals. Our study found that noise and disturbance derived from groups of people wandering through or loitering on the street, from deliveries and services, from queues outside cafés and bars and from car horns. It was not derived from the actions of a small number of excessively anti-social individuals, but came from large numbers of people going about the business of having a good time. Hence traffic congestion was as severe at 3 a.m. as at 3 p.m., and the noise recorded outside a nightclub at 1 a.m. reached a peak that was equivalent to that of a pneumatic drill. The video recorded by the authors (Figure 4.3) showed refuse vehicles in the early hours of the morning, litter building up and, above all, significant crowds of people passing through the street, as many as 20,000 passing one corner in a 24-hour period. That the overwhelming majority of these people were drunk exacerbated the problem. One long-term resident of the area commented:

> I walked home on Sunday from Charing Cross, it was only 8.30pm, but I felt unsafe because of the crowds of very drunk people, wholly unaware of the space around them and totally uncaring of other people's space – shouting and screaming. I found it threatening. Police warnings of muggings by cash machines; a scene of squalor, degradation and dirt. People, late at night, just pee and vomit all over the place. Thursday, Friday and Saturday nights are the worst.
>
> (Representative, Soho Society, London, interview 2002)

Residents in Covent Garden also experienced similar problems. A leading member of the Community Association commented that:

> The number of people using the area has increased dramatically. When I moved to Covent Garden [10 years prior to the interview] I accepted that there was nightlife, but the balance between the day and night-time economies has changed dramatically. Uses start catering for people at night but they are closed in the day, which changes the sociality and feel of the area.... People with children and regular, conventional jobs have moved away. In their place you have young professionals moving in, who can afford to live here and pay their way in; they don't benefit from subsidized housing, they live there out of choice. There's a high turnover of residents. You get people moving in saying 'This is great, can I join your associ-ation? Who are my neighbours?' But within 6 months or a year they're gone and, when you bump into them or correspond, you say, 'What happened? Why?' They answer, 'We've moved on, we couldn't stand it'. They hate the buzz after a while.
>
> (Representative, Covent Garden Community Association 2002)

Westminster City Council and the Metropolitan Police have since taken action to control the numbers of licensed premises in London's West End and to implement a series of management measures to alleviate the problems described in these residents' accounts. These initiatives and the approaches to management of other cities will be discussed in Chapter 8. That the West End of London should have experienced problems with an expansion of its night-time economy is not surprising, given London's ambition to be not only *a* world city, but *the* world city. Nevertheless, echoing Shakespeare, while some cities seek a 'party city' designation, others have it 'thrust upon them'. It is to the latter category that we shall turn to next.

Edinburgh

The popularity of Edinburgh as a party city owes to a number of factors. As well as having two UNESCO World Heritage sites, Edinburgh is well served for the evening and night-time economy. In 2007, the number of off-licenses in Edinburgh was 638. For a population of 448,624 (2001 Census), this may not appear excessively high, but Glasgow, which has a population of 578,790, has only 442. Edinburgh also plays host to 733 public houses (compared to 709 in Glasgow) and 21 nightclubs. Like London, Manchester or Leeds, Edinburgh has a large city-based population. It is also caught in a similar tension of attracting tourism and business on the one hand, while managing the obverse effects of the night-time economy on the other.

Considering each of these points in turn, Edinburgh's city centre ward includes two principle entertainment sites: George Street to the north, and Grassmarket and Cowgate to the south. As of 2001, this central ward accommodated a total population of 22,340 people. George Street, located in the World Heritage-listed New Town attracts a more upmarket clientele, while Cowgate and Grassmarket, in the medieval Old Town, plays host to venues popular with students from the nearby University of Edinburgh's campuses and halls. George Street runs parallel to the city's main shopping thoroughfare of Princes Street. Though originally residential, insurance and banking firms moved in to the street in the nineteenth century, in turn building impressively designed structures along the 115 ft wide thoroughfare. The Royal Society of Edinburgh and the Freemasons are located here, as are a number of foreign consulates, banks and financial institutions. In recent years, many luxury-brand stores have also moved to the area. By night, George Street shifts from banking and retail to leisure and entertainment, with former banks and commercial buildings having been transformed into expensive bars, restaurants and nightclubs. Outdoor seating is commonplace, and clubs are licensed until 3 a.m. Grassmarket and Cowgate is to the city's south, in the shadow of Edinburgh Castle. The area is noted for its dark, winding streets, steep stairs and eclectic mix of shops, bars and restaurants. Grassmarket and Cowgate attract a younger clientele than George Street, and unlike George Street, which discourages hen and stag parties, Grassmarket and Cowgate allow such parties to take place.

Despite the contrasting clientele and market of each area, like Soho in London, many properties in the central city ward are old and sometimes listed. The usual options for limiting noise from the street below – such as installing double glazing – are not permitted, and tension between residents and revellers has become common. In the Old Town, in particular, resident groups have formed to object to licensing applications and to lobby the council:

> The students who go to drink in the Grassmarket trail back here late at night. They stop off, buy food, drop litter, make a noise, create a, not always a very bad disturbance, but they do create a disturbance. They also knock into shop fronts, which can be quite damaging. Quite often on a Sunday or Monday morning we'll see broken windows, door windows, where they've not smashed for entry but have caused damage because they're lurching. If there's a crowd on the pavement lurching along, suddenly you have to go somewhere! They have to stop traffic from going through there because of drunk people wandering onto the road. They had some incidents. Instead of dealing with the drunkenness, they decided to change the safety routine. There are flats in Edinburgh above all our central areas – there are people living above these pubs! At times it's very distressing.
>
> (Local resident, interview 2007)

The Grassmarket has now become part of what has been termed a 'sensitive area' or 'over-provision zone'. Since 2002, no new licensed venues have been allowed in Grassmarket, Cowgate, Victoria Street, Forrest Road, Candlemaker Road or Bristol Place – unless the council decides there are 'special' circumstances. This has been fortified by Edinburgh's City Local Plan, which stipulates that:

> The change of use of a shop unit or other premises to a licensed or unlicensed restaurant, café, pub, or shop selling hot food for consumption off the premises (hot food takeaway) will not be permitted: if likely to lead to an unacceptable increase in noise, disturbance, on-street activity or anti-social behaviour to the detriment of living conditions for nearby residents or in an area where there is considered to be an excessive concentration of such uses to the detriment of living conditions for nearby residents.
>
> (Homes 2007: 107)

How successful this policy has been is debatable, however, and as minutes from the Old Town Community Council attest, there are still conflicts between licensees and residents, and concerns that new licences are being approved. For the resident spokesperson interviewed, the over-provision zone has helped to ease problems in Grassmarket, but problems remain. For the council, the definition of 'special' is more difficult. Given that there is no provision for capacity within this notion of special circumstances, as noted by a licensing officer, it is difficult to prove that '32 [venues] is fine, 33 is too many'.

The differences between George Street and Cowgate/Grassmarket are most apparent in terms of their physical geography. While George Street is a wide street with broad pavements, and easily accommodates the outdoor seating and crowds, as well as easy access to transport and other services, Cowgate and Grassmarket are based upon a medieval, and hilly, street layout. In design terms, it is in many respects the antithesis of an ideal late-night hotspot. Winding stairs and dark alleys not only pose issues relating to safety and disorder, but access for essential services is limited. Beyond this, however, both areas, at night, are devoted to entertainment and while they attract a different demographic, a licensing officer noted: 'People come to the Licensing Board and say "We're opening an upmarket, prestige place on George Street." You wouldn't get a glass of wine there for under £7. But it doesn't make any difference – they still knock it back' (Licensing officer, interview 2007).

Stag and hen parties

A common feature of the party city is the presence of stag and hen parties. The male 'stag' party (bachelor party in North America) has been a common social ritual in British culture for much of the twentieth century. It is an event that is performed across each end of the social spectrum with a common theme of ritual humiliation, raucous behaviour and alcoholic excess. The recent expansion in the night-time economy and the advent of greater mobility through cheap off-peak rail fares and low-cost air travel has assisted in its amplification and extension. Commercial travel operators have realized its potential and provide packages for entire weekends. Instead of occupying the night before the wedding when guests and relatives are assembled together, the stag party may take place weeks or even months before and extend across an entire weekend or longer. It need not necessarily be held in the same physical location as the wedding either, but may be held in another city or abroad. An industry survey in 2003 estimated that 70 per cent of UK stag and hen parties are now held abroad (Mintel 2003).

The female version, the 'hen' party, has also become more prominent. Recent academic work in North America traces the emergence of 'hen' or 'bachelorette' parties to the 1960s, when they were associated with a bridal 'shower' or the giving of gifts to the bride (Montemurro 2003; Tye and Powers 1998). In North America, as in the UK, hen parties have evolved to incorporate variants of the alcoholic and behavioural excess traditionally associated with stag parties. Together they have become a significant part of the weekend or city-break segment of the tourism and travel industry. One source estimate their total worth to the UK in 2005 as £430 million, and participants can spend up to £400 (US$614) each on a typical weekend (InsureandGo 2006).

Many hen and stag parties make themselves visible in the streets by donning matching t-shirts, sashes and 'novelty head gear' (Anonymous 2006), such as a veil for the bride and stag's antlers for the man. These markers help in the identification of stag and hen parties as perpetrators of behaviour that can move beyond the raucous and embarrassing, such as drinking spirits out of penis-shaped glasses inside a nightclub, to actions liable to offend public decency, particularly in cities that have previously not experienced the excesses of nightlife. For example, a news item described the embarrassment of a father whose bridegroom son was sentenced for two months in jail for stripping naked and jumping into a public fountain in front of the American Embassy in Bratislava, Slovakia. The jail sentence extended beyond the wedding, which had been pre-booked and would cost £20,000 (US$30,700) (BBC News 2007). This was an extreme case, but the frequency of problems arising from hen and stag parties abroad has led the British government's Foreign and Common-wealth Office to charge tourists who get into trouble £85 (US$130) per hour for Embassy assistance (Williamson 2006).

Many cities, such as Edinburgh, have become popular destinations for hen and stag parties. A study conducted by the market research group Mintel, found that the Scottish not only play host to such events, but are also keen stag and hen party attendees. Of 1,979 interviews conducted across Britain, 39 per cent of Scottish respondents had attended such an event. This represented the highest regional average in the UK (Mintel 2003).

Temple Bar in Dublin provides an example of the party-city phenom-ena and the stigmatization that can be attached to this designation. The devel-opment of Temple Bar as a cultural quarter had formerly been documented as an exemplar of good practice (Florida 2002; Montgomery 1995, 2007; Quinn 1996). For readers unfamiliar with the area and the story, Temple Bar is an area close to the centre of Dublin, consisting of a grid of narrow streets enclosed by close-packed terraced houses, interspersed with some fine historic buildings and warehouses. At the beginning of the 1990s, it was suffering from planning blight caused by a proposal to redevelop part of it as a bus garage. As Mont-gomery points out, although many of the buildings were near-derelict, the area housed a lively collection of artists, small businesses and resident entrepre-neurs. In 1992, there were 27 restaurants, 100 shops, about six arts buildings,

16 pubs, two hotels, 200 residents, 70 cultural industry businesses and 80 other businesses (Montgomery 2007: 325).

The area was regenerated through the actions of the business community and residents, with the help of the Irish government and Dublin Corporation. An Act of Parliament established a time-limited development company, Temple Bar Properties Ltd, and the creation of an urban design framework plan set the scene for a mixed-use cultural quarter. Several cultural institutions, most notably the Irish Film Centre, helped to 'anchor' the cultural content of the area. The development company was also empowered to cross-subsidize from within its portfolio and to allow cultural organizations below-market rents. The achievements of the first phase of the regeneration were remarkable, and the area flourished with new businesses, performances, shops and visitors. By 1996, there were five hotels, 200 shops, 40 restaurants, 12 cultural centres and an expanding resident population that was to reach 1,500 by the year 2000 (Montgomery 2007: 325).

At first, the emergence of new bars and restaurants combined with new external performance spaces was encouraging. The press officer for Temple Bar remarked that at that time the 'sight of people sitting on street corners and drinking beer was wonderfully positive' (interview 2002), because it was a sign that people were returning to the area and increasing the viability of its retail and commercial uses. As Montgomery (2007) has remarked, the establishment of Temple Bar Properties Ltd was a risk and had there not been an upswing in the Irish economy, the trajectory of Temple Bar could have taken a different turn. Perhaps the subsequent increase in alcohol-related entertainment uses was in fact a symptom of the precariousness of the area's regeneration. By 2000 there were 44 restaurants, 28 pubs and bars, 15 nightclubs and 12 hotels (McDonald 2000). Their density within the 50 hectare area of the cultural quarter is illustrated in Figure 4.4.

By 1998, Temple Bar had acquired a reputation, not only as a cultural centre, but as a destination for hen and stag parties (Reynolds 1999).

> Ryan Air was promoting low cost trips and large parties regularly arrived for weekends. Cheap flights, sometimes with the hotel thrown in and visitors from the UK get 20p on every pound they spend. So, it's a cheap weekend. Big groups were coming in, singing and dancing in the street, urinating against the walls. The pubs get them in, get them drink, get them out. One pub closes, they go to the one that's open until 2.00am.... If you want to see the real Temple Bar, come in at 12.30 or 1.00am when the pubs are closing and you get people eating in the street, throwing their wrappers around and urinating.
>
> (Residents' Representative, Crampton Buildings, Temple Bar)

4.4

Temple Bar, Dublin 2001. Licensed premises (pubs, clubs, restaurants and bars) are shaded black.

Source: Ordnance Survey Ireland Permit No. 8534 © Ordnance Survey Ireland and/Government of Ireland and authors.

75

By 1998, Temple Bar Properties Ltd became concerned that the image of Temple Bar was becoming detrimental to the area's development. Pressure was coming from residents, one of whom was Frank McDonald, the environmental correspondent of the *Irish Times*. In an article for the paper he lamented 'the spectacle of roaring-drunk English type stag party types on Temple Bar Square' (McDonald 1998). An article in the *Irish Independent* the following year described Temple Bar as 'sordid and sleazy' (Anonymous 1999). Temple Bar Properties Ltd subsequently commissioned a market research company, INDECON, also based in Temple Bar, to examine the sustainability of tourism in Dublin. They concluded that while stag and hen tourism represented only 7.2 per cent of the Dublin tourist market, it dissuaded 13 per cent of potential visitors, thereby losing approximately £57 million per annum of Dublin's economy. Furthermore, it also skewed perceptions of the visitor group to Temple Bar through its dominance. A subsequent study found that only 19 per cent of visitors to Temple Bar were British, the majority being Irish (FM-News 1999).

Temple Bar Properties Ltd, Dublin Corporation and businesses in Temple Bar took action to limit the number of hen and stag parties and to redress the damage to the image of the area. The businesses formed a traders group, Traders in the Area Supporting the Cultural Quarter (TASCQ) and sought voluntary agreements from hotels, nightclubs and bars to not take bookings from large single-sex groups. Dublin Corporation in its City Development Plan of 1999 sought to limit the granting of planning permission for new licensed premises and extensions of those currently existing. They also closed one of the planning and licensing loopholes that had exacerbated the problem, whereby hotels had extended their own bars to the extent that they were operating as de facto nightclubs. Hotels were required to put their drinking areas on the first floor or above and to make its area proportionate to the number of guests staying in the hotel.

Stag and hen parties can be read as an extreme manifestation of the type of behaviour 'licensed' in the mind of its participants through the consumption of large amounts of alcohol in a carnivalesque atmosphere. In the case of Temple Bar, such excesses were unforeseen by the businesses and professionals whose vision led the initial transformation of the area into a cultural quarter. The period in which the neighbourhood changed in its night-time uses from a vibrant cultural quarter into a 'drinking den' appears to have only been three or four years. This suggests that policy-makers and planners have to be alert to the nature of the night-time economy and to understand its dynamism in greater depth.

So far, the negative impacts of an unplanned expansion of nightlife have been discussed as though there was an exact correlation between the opening of a large bar and anti-social behaviour. The correlation is not, however, exact. Other factors have an influence, including the perception of the city itself and the type of behaviour that is anticipated and tolerated. The example of Helsinki provides an alternative narrative of the party city where, rather than allowing for unstructured growth, the city has sought to create a vibrant night-time economy that works for both residents and party-goers.

Helsinki: planned liminal spaces

Helsinki is the northernmost capital city within the European Union. It is a relatively small city with a population of approximately 500,000, with double that number in the wider metropolitan region. Situated at the south of the country, Helsinki is linked to other European capitals by frequent ferry services which connect it to Stockholm, Talinn and Copenhagen. The nineteenth-century neo-classical architecture of the grid-like city centre has a similar grandeur to districts in other northern capitals, such as the Dorotheenstadt in Berlin or the Nove Mesto in Prague. Finland enjoys a high per capita income relative to other EU countries, with a smaller differential between rich and poor than in the UK. In common with other urban centres, the city has been developing tourism. It is popular with older, more affluent visitors as well as business tourists.

Evening activities and nightlife in Helsinki have expanded over the last 15 years. Superficially, this change is surprising. Because of its location, Finland experiences long, cold winters and short, sunny summers. In the popular imagination, there is an association with excessive consumption of alcohol and the climate:

> I suppose, I'm not sure, this is not scientific, I'm not a scientist but I think it's got something to do with this geographical position, where we live. Where the conditions of life are very hard, especially in the wintertime, the day time is – the more up north you go the less you see the sun, which means it's very sort of depressing. It's always black, no sunshine, it's cold, you have to survive under those conditions. I think alcohol is one of those things that makes life easier. It's always been part of the Finnish culture; people entertain themselves with alcohol.
>
> (Director, Business Development, Office of Economic Development)

Throughout the greater part of the twentieth century, alcohol consumption in Finland was subject to strict regulation. After a period of state-imposed temperance from 1919 to 1932, from 1932 to 1995 a government organization, Alko, controlled the production, trade and wholesale and retail trade of alcoholic drinks. Control was tight, but by the 1990s there were some relaxations, such as grocery chains being able to sell drink in addition to the state outlets and an

Alko-controlled licensing scheme for restaurants and bars (Alavaikko and Oster-berg 2000). In the 1970s, opening hours were highly constrained, there were few nightclubs and no street cafés. These restrictions began to loosen in the 1980s and restaurants allowed customers to drink outside their premises. The first major change occurred when Finland joined the EU in 1995 and the state monopoly was broken up, although it is still only possible to make off-retail pur-chases of strong spirits from Alko-controlled stores. Beer, cider and alcopops are sold in most grocery stores, petrol stations and kiosks.

Licensing is controlled by the regional state; South Finland in the case of Helsinki. Municipal authorities have an input into hours of operation, as well as controlling land-use, policing and the public realm generally. Licensed premises are able to sell alcoholic drinks between the hours of 9 a.m. and 4 a.m. and to sell weak beer from 7 a.m. to 9 a.m. Commercial de-regulation of the state monopoly facilitated a growth in the numbers of cafés, bars and clubs. In Finland as a whole, the number of on-licenses increased from 3,500 to 5,300 from 1995 to 2003 (Törrönen and Karlsson 2005: 97). Helsinki has a large number of restaurants, standing at 850–900 in 2007. Most of these are concen-trated in the small, historic, commercial centre of Helsinki, but there are major clubs and small concentrations elsewhere in the city, such as Kallio and Punavuori.

This sudden expansion and liberalization led to changes in the use of public space. Public intoxication was a legal offence until 1969 and public drinking was banned until liberalization occurred in 1995. Between 1995 and 2003 public debate in the press about the impacts of alcohol reform focused on a rise in anti-social behaviour in public spaces and, in particular, the impacts of public urination, noise and incivility. The issue of youth drinking formed a target for concern. Local municipalities made their own by-laws and the City of Hel-sinki banned drinking in public places. In 2003, the Finnish Parliament passed legislation to ban drinking in public spaces, with the exception of parks, which were defined as natural areas, not city streets or squares (Törrönen and Karls-son 2005).

Helsinki has a reputation as a 'safe and smart' city in terms of night-life. Interviews with representatives from the City Council, academia and a business organization attributed this to a number of different factors. Perhaps the most important of these was the high price of all kinds of alcoholic drinks, higher than in non-Scandinavian countries and a result of nationally imposed taxation. This difference has been reduced since 2004, when the Finnish gov-ernment was obliged to reduce the taxes on alcohol. Other suggested controls included the weather (it is so severe in winter that external revelry is positively dangerous), visible policing and the historic legacy of a strong social demo-cracy. A pan-European study of crime has pointed out there is a paradox, however, because official figures suggest that the crime rates per 1,000 of the population in Finland as a whole are amongst the highest in the EU. This apparent contradiction can apparently be explained by a greater willingness to report crime and transparency in public processes (Van den Berg *et al.* 2006). In terms of the problems caused by drunkenness and low-level anti-social behaviour, some friction was reported between the residents of Kallio, which is a working-class district and the disturbances caused by customers evacuating from bars and clubs. This appears to have been resolved by the regional state pulling the terminal hour back to 2 a.m.

Public policy in the decade following liberalization was directed towards promoting Helsinki for cultural and business tourism. The cultural and creative sector is large relative to the size of the city. The Director of Cultural Policy explained that one-third of jobs in Helsinki were in the creative sector or in tourism, which translates as 33,000 jobs for 2007. This figure had grown from 28,000 in 2005. Active in the culture and creative sectors were 4,800 private companies. A representative from the City Tourist office explained that 'Helsinki is more for cultural things, shopping and everything that goes along that'. Accordingly, the City Council encourages visitors towards its design district and its restaurants, and runs marketing campaigns such as the 'Nordic Oddity', which highlighted the bohemian and melancholic aspects of the Finnish national character.

Helsinki's nightlife is supported, as in other cities, by the large numbers of students and by the presence of a high proportion of single and two-person households. There are a handful of nightclubs, each with capacities between 1,000 and 2,000. There are also a number of smaller clubs and promoters who hire out restaurants and warehouse spaces. Different subgenres of music have appeared, as have other subcultures. The largest gay bar, for example, has a capacity of 1,500. Although the major clubs are in single ownership and there are restaurant and pub chains, there are a large number of independent owners and promoters in each sector. The Cultural Department of the City Council now takes the view that Helsinki has been transformed into a city that is lively at night:

> In Helsinki ... we are proud we get very high credits for having a lively city also after dark and also on other days, on Friday and Saturday too. I've been surprised to go to several European cities now from my small Helsinki, relatively small, 1 million people, and see what boring places they are after 6 o'clock.
>
> (Representative, City of Helsinki Cultural Office)

Further liberalization is being pursued in the promotion of festivals. Helsinki hosts two important summer festivals, at the beginning and end of the summer. The 30 April/1 May festival, Vappu, sees people enjoying the long summer night outside in the parks, and the city is kept free of cars. Following a 100-year-old tradition, people stay out late or even all night on the evening of 30 April, drinking heavily. On the morning of 1 May they gather in a park in the south of Helsinki and sing the national anthem, then stay out all day, eating and drinking outside.

The second festival, the 'Night of the Arts', is relatively new and takes place in the city centre on the last Thursday and Friday of August. This is part of a ten-day celebration of the arts and music. On the 'Night of the Arts', museums, galleries, book shops and other art institutions stay open until the small hours. Approximately 100,000 people come out to celebrate both events. The City Council has expanded the number of festivals it hosts, from the more traditional Helsinki day to the Tuska heavy metal festival. The Eurovision Song Contest also provided a new point of departure when the city hosted the event in 2007.

> What was really significant about this process, the whole city was living a week on this. Also very strange things popped up, not related to the Eurovision song contest, people could have any approach they wanted, but enjoying the feel. We had lots of people on the streets, the Friday evening before the final when they discuss the song contest, because it's a week program, we had a free festival, open free festival, we had between 50,000 and 100,000 people out on the streets in three hours.
>
> (Representative, City of Helsinki Cultural Office)

Spurred on by this success, the mayor is keen to expand the numbers of events and festivals. A Helsinki-based group of promoters, club owners, media representatives and politicians have made trips to other European capitals to see how they could develop and expand Helsinki's offer. The City Council officials from relevant departments have developed a system for cleaning up after events, regulating security, providing temporary toilets and so on. The city is encouraging the development of events as part of a boosterist strategy, to support tourism and economic development. Nevertheless, the Tourist Board is clear that while it is keen to disrupt the narrative of Finnish melancholia with its obverse, it is not proposing to attract the new market of stag and hen parties, which is helping to drive the designation of party cities and, in turn, drive the transformation of place from heterotopia to dystopia.

Concluding comments

The expansion of nightlife offers a tantalizing promise to city governments. Growth in nightlife provides new jobs, opportunities for training and the prospect of attracting inward investment. Nightlife boosts drink- and food-related entertainment, fashion and live music. Supply chains for each of these are augmented. Small entrepreneurs can join in and, if successful, expand their businesses. There are place-based benefits as well, since some of the most creative forms of new clubs and bars often start life in old, crumbling warehouses, disused industrial sheds and rickety, terraced shop-houses. The growth of nightlife draws young people into urban centres and encourages them to enjoy an intense urban experience in which they collectively effect the transformation of place. At its best, nightlife neighbourhoods and zones encourage a party atmosphere and the prospect of a heightened form of everyday life in which joy and pleasure are tangible and the constraints of everyday behaviour are loosened.

However, because, in the non-Muslim world, legal night-time entertainment tends to revolve around alcohol, it seems that it is easy for the loosening of social boundaries to tip over into threat and aggression. The tensions surrounding drug taking also play a part in terms of buying and selling. Aggressive shouting, fighting and urinating in the street are further transgressions; none of them attractive. Residents who live nearby these manifestations of bad-tempered or violent sociality can find their lives to be unbearable. Ultimately, the negative externalities of nightlife can undermine the positive benefits in economic terms, if businesses, visitors and residents are driven away.

The processes by which a cultural, bohemian quarter can evolve into a sleazy 'drinking den' are little understood. More research is required to comprehend the activities of cultural entrepreneurs and corporate providers and their interaction with the property market. Certainly there is much evidence to suggest that a concentration of alcohol-related entertainment, unregulated and unpoliced, transforms a neighbourhood from a good-natured party space into an urban nightmare. The precise point at which this change over occurs cannot be specified with any precision. In Temple Bar it appears that the collective impact of the expansion of individual premises, with hotel bars turning into major nightclubs, and the advent of the 'superpub' was significant. Similarly, in Soho individual premises expanded and others turned in to café bars cum nightclubs. The monoculture of youthful vertical drinking allied to the extreme of enhanced social rituals, such as hen and stag parties, all play their part. Behind these impacts on place lies another negative effect that will have equally far-reaching implications. This is the impact of the excessive consumption of alcohol. The preceding two chapters have concentrated on the external factors that promote over-consumption. Drunkenness, however, starts with an individual and it is to this that we shall turn to next.

Chapter 5

Binge-drinking Britain?

If you take a 20-something male and a young girl in a very skimpy bikini with a big leather belt around her full of shots [of spirits] going 'come on have a shot, it's only a pound, have one, have one' and all his friends egging him on and then she says 'come on, be a man, have another one' ... These young-sters get wasted on a few cheap shots of rocket fuel ... and yeah, they have a great night.

Licensee, north-east England, 2006

The way I see it is people will come out and they will drink but there is a law that states you're not allowed to serve someone who is drunk, which we stick to. They can only binge to a certain point in my bar.

Licensee, north England, 2006

Introduction

The previous chapters have charted the way in which well-intentioned policies to create a more vibrant late-night urbanism were undermined by the unfore-seen expansion in the numbers, capacities and opening hours of licensed premises. This is not the only issue associated with loosely regulated growth. The implementation of new licensing legislation in England and Wales acted as a catalyst for an intense discussion about the health, crime, social and civil consequences of alcohol consumption (Plant and Plant 2006; Redden 2008). Tabloid newspapers in turn adopted the tag 'Binge Britain' as a catch-all phrase to describe late-night violence in town centres, anti-social behaviour and excessive drinking. Speaking after the launch of the *Alcohol Harm*

Reduction Strategy for England (AHRSE) (Cabinet Office 2004), the former Prime Minister Tony Blair referred to alcohol as 'the new British disease' and proposed working more closely with the alcohol industry to manage its effects. In Australia, the New South Wales Liquor Act 2007 was passed against a comparable backdrop of alarm, particularly in relation to late-night violence and the consumption of alcopops by under-age drinkers. In March 2008, three months before the implementation of New South Wales' new Liquor Act, the Australian Prime Minister Kevin Rudd launched the National Binge Drinking Strategy; a three-tiered approach that included an advertising campaign, a programme to assist young drinkers and a community-led initiative to tackle binge drinking.

Despite these interventions, governments and local councils are caught between dual aims and functions. On the one hand, as was explored in the previous chapter, governments have been involved in measures to encourage the growth of vibrant late-night cities and cultural economies. Increasing tourist numbers and regenerating depressed neighbourhoods is dependent upon allowing for a diverse array of evening and night-time venues and cultural offerings. On the other hand, they are held increasingly accountable for managing the harmful effects of excessive consumption. Whether this is achieved through increasing excise duty on the alcohol industry or increasing late-night policing, it is the state that is held accountable for ensuring a safe and civil society. This has put governments, both locally and nationally, in an increasingly precarious position: liberalizing access and enabling the development of vibrant night-time economies while acting as moral guardians and ensuring public safety. These two aims, encouraging tourism and cultural regeneration while tackling alcohol misuse are not, of course, contradictory. As the debate about binge drinking has played out, however, enabling the growth of the night-time economy has been shrouded by more pressing debates about the health, crime and social impacts of alcohol misuse.

While governments are blamed for lack of regulation, an issue we shall be returning to in the next chapter, the licensed trade is criticized for its greed and its lack of care for the consequences of its growth in sales. Representatives of the trade routinely proffer the view that the British have a natural propensity towards drunkenness, an attribute that defies regulation. There are also divisions between 'on-license' premises, that is pubs, clubs and bars where alcohol is consumed on-site and 'off-license' premises, that is shops that either sell alcohol with other goods, such as supermarkets, or are specialist purveyors, such as general provision 'off-licenses' and wine merchants. As the debate about drinking, drunkenness, regulation and public health has intensified, these divisions have become wider, with the on-trade arguing that increased consumption may be attributed to the off-trade.

Then there is a further quandary, which refers to the extent to which either the state or the licensed trade is at all responsible for individual patterns of alcohol consumption. The freedom of the market, the responsibility of the individual and the role of the state in mediating between the two are, as Nicholls argues, unresolved debates that find precedents in the eighteenth and nineteenth centuries (Nicholls 2006). Outright prohibition may no longer be a valid solution to the 'Binge Britain' crises, but other responses and proposals from the past find accord in contemporary times. Then, as now, there were roughly two responses to managing the problem of excessive consumption: restricting access, be this by higher prices or tougher penalties, or by holding the individual accountable for their own choices and any resulting consequences.

The question of responsibility is a difficult one, and all too often results in the passing of blame between different players or social forces: young drinkers, irresponsible licensees, advertisers, the lack of meaningful jobs, a normalization of drunken behaviour or increased drinking at home. From another perspective, alarm about alcohol consumption in Britain has been rebuked by some commentators as merely an attack on the young (Treadwell 2008) and/or the working-class. In this account, the moral panic that surrounds binge drinking erroneously associates excess consumption with youth culture and detracts from the more serious issue of middle-class consumers drinking at equally harmful levels in the privacy of their own homes. In this account, alcohol misuse is recognized as a problem. However, continuous media attention on youthful binge drinking obscures other factors, other drinkers, and serves merely to further entrench the view that young people are the cause of all forms of anti-social behaviour.

As these arguments suggest, binge drinking is an emotive and complex topic, and alarm over excess consumption goes far beyond a simple concern with the health consequences of alcohol consumption. Instead, a set of other anxieties and concerns are raised: the corporatization of nightlife (Chatterton and Hollands 2003), fear associated with urban centres (Bromley *et al.* 2001), the behaviours and practices of young women (Guise and Gill 2007; Jackson and Tinkler 2007) and young people (Järvinen and Room 2007; Newburn and Shiner 2001; Valentine *et al.* 2008), the power of the market over and above the state (Heather 2006), the relationship between class, free market capitalism and neo-liberalism (Winlow and Hall 2006), the 'respect' agenda, the power of advertisers and the media (Ofcom 2004) and a general unease with the city at night. Over the past decade, these anxieties have coalesced around the figure of the drunk and typically young late-night reveller who is cast as either a victim of boredom, de-industrialization, neo-liberal planning policies and unscrupulous marketers, or as a willing perpetrator responsible for their own behaviour; whatever the consequences.

This chapter will explore the contradictory ways that bingeing is defined and understood. First, we examine several competing definitions of the term. The overall scale of the problem with particular regard to disorder and public health is examined, followed by the debate about private versus public consumption. An exploration of international and historical comparisons are made, and we consider to what extent current responses are conditioned by a 'moral panic'. The chapter ends with proposals for managing binge drinking, such as pricing.

Definition

The definition of what actually constitutes binge drinking is disputed and wavers between measurable quantities, in current terms called 'units' or 'standard drinks', to purely subjective accounts (Measham 2004: 316) such as drinking to the point of inebriation. Binge drinking has also come to be a loosely applied umbrella term to characterize any form of late-night anti-social behaviour in urban centres. Rather than quantifying an individual's blood alcohol content, or measuring how many units they have consumed, bingeing can be used to refer simply to any behaviour involving alcohol that is perceived to be rowdy or threatening (Hobbs 2005b: 24).

In the UK, binge drinking is commonly defined as consuming more than eight units in a single session for men, or six units in a single session for women. This figure is based upon a doubling of the actual recommended daily limit, which is currently set at 3–4 units for men and 2–3 units for women. In order to reach his daily limit, a man would need to consume two pints of beer or 3–4 small glasses of wine, while women would only need one-and-a-half pints of beer or 2–3 small glasses of wine. With these representing the recommended daily limit, binge drinking is thus defined as a doubling of these measures.

There are, however, several areas of dispute. The General Household Survey, an ongoing survey carried out by the Office for National Statistics, defines bingeing as *more* than eight units for men and more than six for women. The Health Survey for England, on the other hand, proposes eight units or more for men and six units or more for women. Whether binge drinking is understood as consuming exactly eight units, or whether counting starts at more than eight (meaning counting effectively starts at nine), has implications for the number of people said to be binge drinking (Berridge *et al.* 2007: 21). To add a further layer of complexity, the system of measurement itself is not altogether clear. To begin with, contemporary drink measures or the alcohol content of many alcoholic beverages is rarely accounted for. A standard bottle of wine, especially 'new world' wine, can have 13–14 per cent ABV (alcohol by volume), while beer and cider may well have an alcohol content of 6 per cent. With the official unit of measurement based on wine containing 9 per cent ABV, and beer or cider 3.5 per cent, this clearly poses problems with assessment. Moreover, exactly what constitutes one unit, that being a 125 ml glass of wine, for example, may not always be obvious to consumers, especially if they happen to be 'free pouring' at home or drinking out of today's standard larger glasses.

To speak of binge drinking in daily terms, more than eight units for example, is also slightly misleading and is only a recent occurrence. Prior to the mid-1990s alcohol consumption was more generally understood in weekly terms, with the recommended limit set at 21 units a week for men and 14 units for women. The problem with this method was that a man could conceivably consume 20 units in a day and still be under safe weekly limits.

For the Institute of Alcohol Studies' *Binge Drinking* report, more than five standard drinks in one session constitutes bingeing (Institute of Alcohol Studies 2007: 9). This method is based upon Wechsler's research into college drinking patterns in the United States. Wechsler is attributed by Berridge with having initially popularized the term 'binge drinking' in the 1990s (Berridge *et al.* 2007: 18). While he popularized the term, his use of the five standard drinks system of measurement was taken from the earlier *Monitoring the Future* study conducted by the University of Michigan. Since this was a US-based study, and thus based on the US definition of a 'standard drink', it is not clear how the five drinks model translates into the British metric system.

An alternative unit-based system comes from the United States, where the National Institute of Alcohol and Alcoholism (NIAAA) defines binge drinking as 'a pattern of drinking alcohol that brings the blood alcohol concentration (BAC) to 0.8 g/L or above' (Department of Health and Human Services 2004: 3). Then there is the Australian model, which is not based on units either, but rather 'standard drinks', with one standard drink containing 10 g of alcohol, or a 30 ml nip of 40 per cent spirits or 375 ml of 3.5 per cent beer.

Every unit-based model is potentially flawed in that eight units consumed over an entire day alongside food results in different consequences than eight units in a single session over a short period of time (Institute of Alcohol Studies 2007: 1). Putting this into a broader context, as Berridge's study into the normalizing of binge drinking points out, unit-based definitions fail to consider context, how quickly one consumes, or body weight, which will each affect the rate of intoxication (Berridge *et al.* 2007: 19). The usefulness of the unit-based system of assessment was further questioned by a newspaper report claiming the unit model was not based upon an actual scientific study. Richard Smith, former editor of the *British Medical Journal* and a member of the Royal College of Physicians' working party which recommended safe limits in 1987, disclosed that the guideline of 21 units per week for men was 'plucked out of the air. They weren't really based on any firm evidence at all. It was a sort of intelligent guess by a committee' (Norfolk 2007). Smith has since responded to this statement in an article published in the *Guardian* where he noted that the figures *were* based on scientific experience, if not one particular scientific study (Smith 2007b).

Given the complexity of the unit-based system, subjective measures may offer an alternative to assessing binge drinking. A subjective measurement would begin from the principle that bingeing is drinking for the purpose of inebriation. Another method would be based upon self-disclosure of, say, having been 'very drunk' once a month or more (Measham 2004: 316). The perception of 'being very drunk' may differ remarkably for different people, and will depend on one's own alcohol tolerance, but a subjective measurement of bingeing at least relates specifically to actual intoxication. Further methods of calculation which are not specifically based upon units, but instead context, rate or outcome, include consuming more than one alcoholic drink per hour, drinking for more than three days in a row, or drinking to the point of collapse. The government's 2007 report *Safe. Sensible. Social.*, a follow up to AHRSE, follows this method and defines binge drinking as simply 'drinking too much alcohol over a short period of time, e.g. over the course of an evening, and ... drinking that leads to drunkenness' (HM Government 2007: 3). Of course, it is important to bear in mind that whether based on clinical units or otherwise, any method that relies on self-disclosure has inherent difficulties (Midanik 2006). Young people in particular may over-emphasize their capacity for alcohol, sometimes with amusing results (Plant and Plant 2006: 28).

While subjective methods may appear to lack rigour, they are useful in terms of measuring a drinker's motivations. Discrepancies in how binge drinking is currently defined, in a quantitative sense, not only pose problems for researchers seeking to draw historical or cross-cultural comparisons, it makes it exceedingly difficult to formulate preventative measures (Berridge *et al.* 2007: 22). A more adaptable model could begin from the point that any drinking where the intention is to become intoxicated can lead to social or health problems. It is also worth highlighting that some individuals are highly sensitive to alcohol, and may display classic signs of over-consumption after only one or two 'units'. While they may not have been bingeing according to a quantifiable system of measurement, if their intention was an example of 'determined drunkenness' (Measham 2006), and that intention was achieved, then it requires us to examine qualitatively the reckonings behind this desire. Of course, on these grounds the obverse could equally be true. A consumer may not experience drunkenness after eight or more units, and their intention may never have been to achieve a state of drunkenness. However, the health impacts of their binge drinking would be of concern. In this example, a unit-based model would be more effective.

More recently, the definition of problem drinking has been extended to two new drinking categories, hazardous and harmful. With bingeing strongly associated with late-night disorder and young consumers, this latest method explains the variations in excessive drinking practices and adds precision to the over-burdened term of bingeing. Researchers from the North West Public Health Observatory at Liverpool John Moores University, define hazardous drinking as consuming 22–50 units per week for men and 15–35 units per week for women. Harmful drinking is classed as regularly consuming more than 35 units per week for women, and over 50 units per week for men. Again, this is a unit-based model. However, it adds a further layer of nuance to this troubling debate. To summarize the argument thus far, there is a good deal of dispute about what constitutes binge drinking. There are wider areas of conflict over systems of measurement, out-dated systems for measuring alcohol content and a lack of transparency in exactly how established guidelines were formulated. Qualitative methods may, in turn, reveal more about consumer intentions, and allow for a more precise understanding of the reasons why people drink to excess. This may be of little use when it comes to assessing Britain's overall levels of consumption, or making historical and international comparisons, but one's intentions and the context in which consumption occurs can reveal a great deal about our national and historical attitudes to alcohol consumption.

Scale and severity of the problem

Though defining binge drinking has proven to be difficult, levels of consumption in the UK were, at least until 2006, believed to be on the increase. Figures are not always clear, and can even be contradictory. However, it is safe to say that consumption by British women and the under-age has caused particular alarm. AHRSE takes an initially positive view, stating that 'alcohol plays an important part in our society and makes a substantial contribution to the UK economy' (Cabinet Office 2004: 1). The report goes on to further point out that 'over half the adult population drinks less than 14/21 units a week'. While, as discussed earlier, there are problems with weekly measurements in that they do not reveal the amount consumed in a single day, the strategy does go on to examine the negative side of the nation's drinking habits: one in five women, and one in three men consume more than the recommended 14 and 21 units per week.

The Information Centre, a unit within the Office for National Statistics, came to a similar conclusion. Their 2007 report, *Statistics on Alcohol: England 2007*, is based on research compiled using data from the Home Office, the General Household Survey, the Department for Transport and Her

Majesty's Revenue and Customs England (Office of National Statistics 2007b). It found that in 2005 73 per cent of men and 58 per cent of women had consumed alcohol at least once in the week prior to interview. For 13 per cent of men and 8 per cent of women, consumption had occurred every day in the previous week, and 34 per cent of men and 20 per cent of women had exceeded their recommended daily limit in the previous week. In terms of actual binge drinking, defined in this report as *more* than twice the recommended daily limit, 24 per cent of men and 14 per cent of women had binged. Figures from the 2005 General Household Survey, reproduced here in Table 5.1, paint a comparable picture.

From a more positive angle, using the 2007 Information Centre figures, the majority of people do not exceed safe limits. The pervasive myth of Britain as a 'drink sodden' nation is incorrect in that 76 per cent of men and 86 per cent of women are not binge drinkers. Be this as it may, what these figures do not reveal is the sharp increase in alcohol consumption that has occurred over the past decade. Elsewhere in Europe, consumption appears to be dropping, though, as is discussed later, this too has been disputed. In the United Kingdom consumption appeared to be on the increase. An Interim Analytical Report, ordered by the government in preparation for AHRSE, concluded that: 'If present trends continue, the UK would rise near the top of the [European] consumption league table within the next 10 years' (Prime Minister's Strategy Unit 2003: 18). As Heather notes, at least until 2006, this prediction appeared to be well on course (Heather 2006; Prime Minister's Strategy Unit 2003: 21).

Since 1951, consumption in Britain has increased by 151 per cent (Prime Minister's Strategy Unit 2003). Much of the focus around binge drinking has fallen on young people, and this has not been entirely misplaced. The DCFS' Youth Alcohol Action Plan claims that while the number of young under-age consumers drinking regularly has dropped since 2000, those that do drink are consuming over 50 per cent more. In 1990, for example, 11–15-year-olds consumed five units per week. Based on figures from the British Medical Association's *Alcohol Misuse: Tackling the UK Epidemic* (2008), the average number of units consumed by 11–15-year-olds has now risen to 11. In terms of gender

Table 5.1 **Binge drinking by age**

Age	16–24	25–44	45–64	65+	All
Men					
More than four units and up to eight units	12%	18%	20%	12%	17%
More than eight units	37%	28%	17%	4%	21%
Women					
More than three units but up to six units	18%	17%	12%	3%	13%
More than six units	23%	11%	4%	1%	8%

Source: Office for National Statistics General Household Survey, 2005, Crown Copyright.

differences, 37 per cent of 15-year-old boys interviewed had drunk at least once in the previous week. For girls, the figure was 47 per cent, setting forth a flurry of news reports about the number of teenage girls binge drinking. Of particular concern, for both boys and girls, is evidence of 'determined drunkenness' whereby 35 per cent of 11–15-year-olds who had consumed alcohol in the previous four weeks had done so purely to get drunk. For 15–16-year-olds, the figure rose to 56 per cent having drunk heavily in the month prior to interview (Department for Children, Schools and Families *et al.* 2008: 8).

In contrast to the perception that young people have the monopoly on all forms of alcohol consumption, it is older people who are more likely to drink regularly. As of 2005, 28 per cent of men and 18 per cent of women aged between 45 and 64 consumed alcohol on five or more days in the week prior to interview. This compares to only 10 per cent of men and 5 per cent of women aged 16–24. Differences emerge, however, in the actual volume consumed, with 42 per cent of men and 36 per cent of women in this age group over the recommendation limits, while only 16 per cent of men and 4 per cent of women aged 65 and over did so (Office of National Statistics 2007b: 4). By the age of 15, one-third of young people in England drink weekly, with 41 per cent drinking for the purpose of intoxication (Hughes *et al.* 2007: 1).

Needless to say, figures proclaiming the extent of binge drinking have been challenged. Datamonitor, a market analysis group, found that the number of units consumed by British drinkers had decreased by 0.5 since 2000, and though Britain still tops the European average, the gap is narrowing.

Table 5.2 **Average weekly alcohol consumption (units), 1998–2005**

Age	1998	2000	2001	2002	2005
Men					
16–24	25.2	25.9	24.8	21.5	18.2
25–44	17.1	17.7	18.4	18.7	16.2
45–64	17.4	16.8	16.1	17.5	17.7
65+	10.6	11.0	10.8	10.7	10.4
Total	17.1	17.4	17.2	17.2	15.8
Women					
16–24	11.0	12.6	14.1	14.1	10.9
25–44	7.1	8.1	8.3	8.4	7.1
45–64	6.4	6.2	6.8	6.7	6.3
65+	3.2	3.5	3.6	3.8	3.5
Total	6.5	7.1	7.5	7.6	6.5
All					
16–24	18.0	19.3	19.4	17.6	14.3
25–44	12.0	12.9	13.3	13.3	11.3
45–64	11.7	11.4	11.3	11.9	11.7
65+	6.3	6.7	6.6	6.8	6.5
Total	11.5	12.0	12.1	12.1	10.8

Source: Office for National Statistics General Household Survey, 2006, Crown Copyright.

On a typical night out, drinkers consume 6.3 units. This is higher than the 3.9 units in the Netherlands, but closer to the German (5.5) or Spanish (5.3) units of consumption (BBC News 2006). The picture painted by the Institute of Alcohol Studies also tells a slightly contrary story. The table used in their *Drinking in Great Britain*, reproduced here as Table 5.2, does not point towards an overall increase, but rather a rise then decline in weekly consumption between 1998 and 2005. The figures are taken from the Office for National Statistics General Household Survey (2006) 'Smoking and drinking among adults'.

These figures should be taken with care, however. As noted earlier, binge drinking is the problem at hand and weekly figures do not reveal how much is consumed in a single session. Equally, we know that some people do not drink at all, and as average figures, there are some members of the community drinking a good deal more than others. In summary, the extent of binge drinking in Britain today appears to be on the decline. There are trends, however, particularly concerning people under 18 years of age, that point to excessive consumption. As well as the health concerns, this raises the no-less-difficult issue of the relationship between binge drinking and anti-social behaviour.

Policing and crime

Over-consumption is one component in a string of factors relating to late-night anti-social disorder; too many people competing for too few resources, the elision of macho culture, loyalty, chivalry and drunkenness (Graham and Homel 2008: 71–8) and a concentration of venues and people in a small area. As Hobbs *et al.* state: 'If you cram tens of thousands of individuals together from the age group most prone to criminal behaviour and then fill them with alcohol, does anyone really believe it won't "go off"?' (Hobbs *et al.* 2003: 11).

The relationship between alcohol and disorder is complex, and consuming alcohol is not necessarily a precursor for violent behaviour. Barr (1998) goes so far as to state that there is no evidence to suggest alcohol results in higher levels of aggression at all and refers to a well-cited study from the University of Wisconsin. Conducted in 1974, the study entailed a group of 520 volunteers served either alcohol or tonic water, with both groups told they were drinking vodka and tonic. Alcoholics, even if they were drinking plain tonic, stopped having tremors and men in the non-alcohol group reported feeling more relaxed. Women, on the other hand, even those drinking the tonic, reported feeling more anxious, a response Barr attributes to women having to be more careful of their 'temptations' (Barr 1998: 24). Whether women today would report feeling more anxious is unknown, but what is important is that the study raises a host of questions about how the volunteers' behaviour changed under the impression they were consuming alcohol. As noted by Fox and Marsh, citing the alcohol scholar Dwight B. Heath: 'There is overwhelming historical and cross-cultural evidence that people learn not only how to drink but how to be affected by drink through a process of socialisation.... If behaviour reflects expectations, then a society gets the drunks it deserves' (Fox and Marsh 2006: 18).

Other mitigating factors therefore need to be considered. Pricing, policing, gender, or even the design of venues, as amply demonstrated by Graham and Homel (2008) may, in conjunction with other factors, influence behaviour. Following the responsibilization model, which has gained prominence in recent years, it is easy to discuss bingeing as an individual problem when in fact heavy drinking and/or violence occur within a much broader context. If the experiment recounted earlier remains irrefutable and alcohol does not necessarily equal violence, attention needs to be turned instead to far more complex questions about the reasons behind anti-social and aggressive behaviour. Moving away from a statistic-based study of violence and alcohol, Hobbs' work examines the cultural and economic foundations of violence in the night-time economy. His most well-cited study of 'bouncers', or door security staff, was conducted with colleagues who had worked inside the securities industries studied (Hobbs *et al.* 2003). Two of Hobbs' colleagues, Steve Hall and Simon Winlow, have since released their own study, *Violent Night* (2006), which extends the analysis of violence and the night-time economy in terms of contributory economic, social and political factors. Based on extensive interviews with victims, perpetrators and the police, the study is framed by broader questions about youth culture, class and the effects of a neo-liberal regime. While their study offers little in the way of quantifiable statistics, and therefore whether crime is increasing or reducing, it frames late-night violence, youth culture and binge drinking in light of poor work prospects, community decay and a night-time economy based purely on *economic* gain. They argue that:

> the promise of hedonistic release from the stifling monotony of service work, the pressures of education and career building, the moral restraints of parents and the gloom and isolation of decaying communities has been crucial in the remoulding of weekend leisure to fit in with shifting dynamics and the basic profit-maximising demand of consumer capitalism.
>
> (2006: 103)

Health and economic costs

Alcohol has long been examined and understood in terms of both health benefits and costs (Burnett 1999). In Britain today, this debate has yet to be

resolved, with the mainstream press seemingly rotating stories about the benefits of moderate consumption one day and the dangers alcohol can pose the
next. In terms of the dangers, according to AHRSE, approximately 22,000
deaths per year are a result of alcohol consumption in Britain. This figure is
compounded by findings from 2006 which report that between 1991 and 2006,
the number of deaths attributed to alcohol misuse doubled, from 4,144 to
8,758. The largest increase was for men between 35 and 54 (Office of National
Statistics 2008a).

In economic terms, the cost of alcohol consumption to the British
economy has been further calculated at £20.1 billion. It should be stated that
this figure is not the cost of specifically 'binge drinking', but of alcohol in
general. Health problems account for £1.7 billion of this total figure. Human and
emotional costs stand at £4.7 billion, physical costs at £7.3 billion and lost productivity at £6.4 billion. Health costs may be the lowest figure here, and it is
alarming that of this £1.7 billion, only 6 per cent, approximately £102 million,
represents monies spent on alcohol services. Moreover, while the cost is
clearly staggering, it is worthwhile bearing in mind that the alcohol industry
annually spends almost £800 million on promotions (Pincock 2003: 1126). The
Information Centre's *Statistics on Alcohol: England 2006* claims that approximately 35,600 NHS hospital admissions where the principle diagnosis related
to mental and behavioural disorders were due to alcohol. Of this, 68 per cent
were men. A key measurement of binge drinking is related to cirrhosis of the
liver, which has risen substantially in the UK in recent years. The Information
Centre's *Alcohol Use 2006* report states that liver disease is less common than
mental or behavioural disorders. However, figures for alcoholic liver disease
more than doubled between 1995/2006 and 2004/2005, from 14,350 to 35,393
(Office of National Statistics 2007b: 51). Other health complications associated
with binge drinking include cancers, infertility and, a more recent discovery,
ruptured bladders (Dooldeniya *et al.* 2007). The report, in the *British Medical
Journal*, states low figures, just three in the span of 12 months. However, it
called for medical specialists to be aware of ruptured bladders if women presented with abdominal complaints (Dooldeniya *et al.* 2007: 992–3).

As well as the impact alcohol has on one's own body, there is the
related issue of its effect on family members. While not necessarily related to
health per se, AHRSE notes that between 780,000 and 1.3 million children in
the UK are affected by their parents' alcohol problems. This represents 1 in 11
children, and returns the discussion to a subject flagged earlier: public versus
private consumption.

Public and private drinking

The issue of youth drunkenness raises a complex debate about the consequences of public versus private drinking. Towards the end of 2007, the
North West Public Health Observatory produced a website mapping hazardous
and harmful drinking in England. The report was immediately controversial as it
highlighted patterns of alcohol misuse in seemingly middle-class homes and
areas. Questions were immediately raised about the nanny state and individual
privacy. Writing in the *Sun*, opinion columnist Samantha Wostear questioned
the researchers' intervention into the matter of home drinking, asking 'Did we
taxpayers really fund this garbage?' Wostear went on to call on researchers 'to
leave us alone and concentrate on what REALLY causes misery – drug-induced
crime, youth binge drinking and the violence that stems from both' (Wostear
2007: 24). Across both the tabloids and broadsheets, there was a comparable
indignation at what was perceived to be evidence of an intrusive state.

In Australia, a similar debate was played out, and again, a distinction
was drawn between middle-class consumption in the home and youthful bingeing on the streets, with a veiled defence of middle-class drinking patterns.
Writing in the *Sydney Morning Herald*, Anne Summers wrote:

> If four glasses of wine enjoyed by adults over dinner is now going to
> be labelled binge drinking, we will need a whole new vocabulary to
> describe kids throwing down 24 vodka shots on a night out on the
> town.
>
> (Summers 2008)

Four glasses of wine is, indeed, binge drinking, but the message followed a
similar line to the *Sun*. Following the maxim that an alcoholic is 'anyone who
drinks more than me', binge drinking was rendered as seemingly more worthy
of intervention and discussion when associated with youth rather than the middle-classes.

A more critical response to this debate comes from Holloway *et al.*,
who have intervened in the private/public drinking debate by specifically examining home drinkers (Holloway *et al.* 2008). In their 'Sainsbury's is my Local',
which refers to the large supermarket chain rather an actual local pub, Holloway *et al.* critically engage with the very notion of 'home'. To place this discussion in context, a study of drinking behaviour by the Office for National
Statistics found 70 per cent of respondents had purchased alcohol from a
supermarket in the previous year. A slightly smaller number, 66 per cent, had
purchased alcohol from a bar, and 61 per cent from a restaurant (Office for
National Statistics 2004: 33). Bearing this in mind, the distinction between
home drinking and public drinking is not always clear. In the most immediate

sense, to posit a distinction between the two obscures the many variations that occur within both spaces, as well as the overlaps. The home drinker may be typically represented as the lonely drunk, but home drinking may occur as a result of parties, dinners, events and celebrations, a final nightcap, as well as the controversial practice of pre-loading. In these terms consumption at home may overlap with drinking within other spaces, or indeed other contexts.

Though the debate about middle-class and private drinking ended abruptly in Britain, in 2008 it widened to a more nuanced discussion about pricing, as discussed in more depth later. The lesson to be learnt from the outcome of the North West Public Health Observatory's research, however, was that there is a reluctance by the British press to fully engage with the issue of private drinking, despite their seeming fixation with drinking on the streets of 'Binge Britain'.

The British disease

One of the more persistent excuses for Britain's levels of consumption, at home or otherwise, is that drinking to excess is seemingly in the nation's genes. As discussed, licensing reform in England and Wales was met with a range of responses. In some quarters, the attitude was not so much concern as indifference. The British, or so the argument went, had long had an enthusiasm for consuming alcohol and a change of law would do little to impact upon patterns or rates of consumption. The eighth-century missionary St Boniface, born in Devon but who spent most of his life on the European mainland, has become exceedingly well cited in recent years. He wrote to the Archbishop of Canterbury: 'In your diocese the vice of drunkness is too frequent. This is an evil peculiar to pagans and to our race. Neither the Franks nor the Gauls nor the Lombards nor the Romans nor the Greeks commit it' (Barr 1998: 25; Plant and Plant 2006: 3). Other well-cited 'evidence' of the British preponderance to drink comes from the Tudors. The first Licensing Act was introduced in Britain in 1552 due to political and social unrest. Between 1604 and 1627 a series of other Acts were passed to manage 'public drunkenness' (Plant and Plant 2006: 5). Here, the issue was less about consuming alcohol than drunkenness, and Elizabeth I is famously known to have drunk ale for breakfast, dinner and supper (Burnett 1999: 113). The fact that alternatives such as milk or water were not as readily available as they are today tends to be lost in recounting this tale. Instead, it is more common to reiterate that Elizabeth I drank lager for breakfast as yet further evidence that the British have always drunk.

Several arguments circulate here, and the subject is more complex than it first appears. First, examples from history about excess drinking in Britain have also become remarkably pliant in terms of either justifying or explaining contemporary behaviour. On the one hand, there is the view that bingeing is a pastime rooted in history that can therefore not be changed. This is not only incorrect, it absolves all parties of any responsibility for excess consumption. More importantly, it also ignores the interventions and strategies that have been affected to control alcohol consumption. Finally, it again confuses three separate issues: alcohol consumption, bingeing and anti-social behaviour.

As argued, the ahistorical view of Britain's drinking patterns is not altogether correct. Far from having always been a nation of heavy drinkers, alcohol consumption has risen and declined in Britain over time. With the current concern over binge drinking, and the ahistorical arguments that we are a nation of drinkers as a starting point, Peter Borsay (2007) examines the parallels between the eighteenth-century gin crises and the moral panic about binge drinking today. He argues that the eighteenth-century moral panic about alcohol was equally focused around women, government complacency, urbanism and changing social mores. Equally, and again as with today, the debates were less about alcohol consumption per se, so much as the consumption of alcohol by the wrong sort of people drinking the wrong sort of beverage. As the modern-day equivalent of gin, it is typically assumed that alcopops cater to young women who, by extension, have 'no taste' and evidently drink purely to get drunk. This attitude towards certain types of beverages as the 'wrong type' also finds other parallels in history. As Barr explains, black-cherry brandy was a mix popular with women in the late seventeenth century (Barr 1998: 28). Equally, he explains how cocktails became increasingly popular in the 1920s and were an especially successful way of 'annoying one's elders' (1998: 73).

If the British had a natural predisposition to drink, one would expect drinking figures to remain fairly constant, which they have not. As noted above, the average consumer in the UK was, as of 2001, drinking the equivalent of 8.6 litres of pure alcohol, an increase of 151 per cent in 50 years. As can be seen in Figure 5.1, however, British consumption has increased and decreased since 1900.

Working further into the past, Burnett's *Liquid Pleasures* provides a useful overview of the historical variations that have occurred in terms of British drinking habits. Not only have figures dropped over the past century, they have also increased and decreased at particular times – notably during each of the World Wars and also the inter-war period. Drinking during the day has also become less common as drinking practices, working lives and the types of work commonly undertaken have changed. Other factors, such as the popularity of cafés rather than pubs, should also be recognized, as should other 'attitudinal and behavioural' changes (Measham and Brain 2005: 265). Other factors Measham and Brain point to include the normalization of recreational drug use, the increased alcohol content of some beverages, the articulation of alcohol

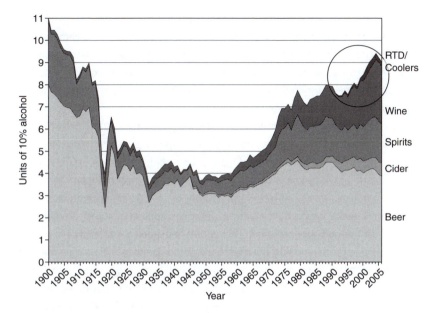

5.1
Alcohol consumption in the UK: 1900–2006.
Source: British Beer and Pub Association, Andrew Tighe.

with sophisticated urban lifestyles and, finally, the promotion of alcohol due to a 'move from "spit and sawdust" working-class back street pubs to modern "chrome and cocktails" city centre café bars with plate glass fronts' (Measham and Brain 2005: 267). Far from an ahistorical and consistent drinking practice, all these factors have, with different intensities, shaped modern consumer drinking behaviours.

International comparisons

The Academy of Medical Sciences' *Calling Time* begins from the simple premise that alcohol consumption in the UK, per capita, has risen 50 per cent since 1970. In the same timeframe, it is argued that consumption in France and Italy has halved. Be this as it may, and despite the recent well-documented rise in consumption, the UK's rates of consumption are relatively moderate compared to other nations in Europe (Heather 2006: 225). Supporting this claim requires comparing international levels of alcohol consumption. While this is possible, it should be remembered that there are variations in measuring alcohol, as well as stark cultural differences in patterns of consumption. In the UK, as discussed earlier, one unit of alcohol is equal to 8 g of pure ethanol. This differs considerably from Japan, where a unit is 19.75 g, Australia and Spain where a unit is 10 g or Italy where it is 12 g. Moreover, when international safe limits are translated into British units, vast differences in safe and unsafe levels materialize. Poland, at 12.5 units per week, represents one of the lowest levels of safe drinking measurements in Europe. This is not only lower than Britain's

21, but substantially lower than Canada (23.75 units), Denmark (31.5 units) or Australia (35 units). Furthermore, countries such as Canada, the Netherlands and Spain do not differentiate between recommended units for men or women. And, since drink sizes do not necessarily match internationally, comparing binge drinking – or any other drinking figures – across nations becomes especially difficult (Berridge *et al.* 2007: 19).

Further to this comparison within Europe, it must be remembered that on a broader scale, Europe, defined here as the European Union, also has the highest rate for alcohol consumption in the world. This is important since it is possible to argue Britain is consuming lower levels of alcohol than the Czech Republic, but this does not mean the country is not still significantly ahead of other, more sober nations. As a whole, European adults consume, on average, 11 litres of pure alcohol per year. This is two-and-a-half times higher than the world average (Anderson and Baumberg 2006: 77). Despite this being a marked decrease from 15 litres, levels in general appear to have steadied, but some nations, such as Ireland, are continuing to rise. Though the figures are difficult to correspond, Ireland and Finland record almost three times the level of binge drinking to Italy, and there is a clear north–south divide (Anderson and Baumberg 2006: 93). Table 5.3 compares pure alcohol consumption, in terms of litres per capita, between 1993–2004. The figures refer to consumption by individuals over 15 years of age.

Though these figures suggest Britain is not at the top of the table for yearly consumption, again we need to emphasize yearly as opposed to daily patterns of consumption. Equally, the figures for Europe-wide patterns of consumption are not entirely transparent. Spain, for example, has lowered their overall consumption since 1975. According to Gual (2006), however, the figures are slightly misleading. The reduction is due to a shift in consuming beer rather

Table 5.3 **Pure alcohol consumption: litres per capita, 1993–2004, ages 15 and over**

	1993	1995	1997	1999	2001	2003
Luxembourg	18.08	17.06	18.47	17.65	17.56	18.00
Czech Republic	15.71	15.83	16.37	16.48	16.21	16.15
Ireland	11.24	11.45	12.81	13.82	14.44	13.47
Austria	13.79	13.41	13.03	12.79	12.25	12.57
Denmark	12.20	12.56	12.51	11.91	11.93	12.08
Spain	11.97	11.35	11.98	11.61	11.43	11.70
UK	9.96	9.71	10.23	10.25	10.73	11.37
Italy	11.11	10.45	9.82	9.36	9.14	10.45
Netherlands	9.75	9.86	9.96	9.91	9.76	9.56
Finland	8.39	8.31	8.56	8.62	8.95	9.31
Greece	11.15	10.55	9.96	9.92	9.30	8.99
Poland	8.35	8.12	8.69	8.43	7.70	8.15
Bulgaria	10.24	9.91	8.58	8.15	7.13	5.89

Source: Office for National Statistics, 'Statistics on Alcohol: England, 2007', citing World Health Organization (WHO): Regional Office for Europe – Health for All database.

than wine, and not necessarily an overall reduction in patterns of consumption. The distinction between drinking wine or beer suggests a layer of debate which concerns the volume of alcohol consumption in terms of pure alcohol, versus actual drinking occasions. Spain may have reduced their overall consumption of alcohol in terms of consuming pure alcohol, but if consumers have simply switched from a high-alcohol beverage such as wine to a lower-alcohol beverage such as beer, the drinking context or culture may not have considerably changed. Spain's reduction of alcohol is commendable if it leads to fewer health problems; however, alcohol is also a social issue and whether we consume vodka, wine or beer, if there is still no alternative to alcohol-centred entertainment then there is less to celebrate.

In Italy, the situation is equally unclear. The Italian Health Ministry reports that 17 per cent of the population binge drink at least once per month and one-fifth of teenagers claim to be drunk regularly (Fraser 2007). Emanuele Scafato, from the Italian Institute for Health, further challenged the belief that there is a substantial gap between British and Italian youth drinking patterns. More notably, comparable reasons for the increase in Italian youth binge drinking were offered.

> Young people no longer drink for enjoyment.... They drink to get drunk. The relationship with alcohol is very different to what it was 10 years ago. We blame the growth in the sale of alcopops, the way the industry encourages young people to drink to be 'cool'. These days you can't be 'part of the gang' in Italy unless you drink. The second problem – is the breakdown of the traditional family unit. Drinking in moderation was something you learned from your father. Young people were encouraged to enjoy a glass of wine at dinner. Now parents work longer hours, the rhythm of life is changing and so is the father son relationship.
>
> (Fraser 2007)

The recent botellón phenomena is another, albeit extreme, example of where locally specific practices need to be considered. Botellón, or big bottle, is the term used to describe the consumption of large bottles, typically filled with wine and cola, in public by young Spaniards (Gual 2006). What began in Andalusia as groups of students drinking take-out in the city's square, developed into a macro-botellón in 2006 when thousands of Spanish youths used the botellón to mix protest with pleasure. In Granada alone, 25,000 drinkers took to the streets in a nationwide event. The botellón has been justified by organizers and attendees on the grounds that licensed venues are too expensive.

These are isolated incidents and there remains divisions between and within northern and southern European drinking contexts, patterns of consumption, and outcome (Anderson and Baumberg 2006). Nonetheless, like alcopops in Australia or Italy and vodka shots in Britain, the Spanish botellón perfectly captures the anxieties that surround binge drinking: young people, urban space and sweet sugary alcohol drinks. These provide the ingredients for a moral panic.

As important as questions about health and anti-social behaviour are, the binge-drinking panic as played out in the press is a confluence of anxieties about the working-class, civility, youth and a general abandonment of the values of work, morality, respect and urban sobriety. These anxieties are modified depending upon the targeted group, such that alarm about women's drinking, for example, is typically about their health and 'femininity' (Day *et al.* 2004). Such is the pervasiveness of the term, however, it is now easy to dismiss genuine concerns about youth alcohol consumption as merely a 'moral panic'. That is, a moral panic concept that rests on a presumed distinction between a media-fuelled debate divorced from reality and the 'real world' – as if the two can be so readily separated. How useful the notion of a moral panic is for exploring binge drinking is also questionable. The term was originally coined by Stan Cohen in his influential *Folk Devils and Moral Panics* (1972). The term was developed further by Marxist media theorists who argued that moral panics, be it about 'trade unionists, homosexuals, teachers, blacks, foreigners' (Fowler 1991: 53) were designed to draw attention away from broader structural issues. This certainly applies to the debate about binge drinking today, where the poor provision of infrastructure, neo-liberal policies and the over-concentration of premises are obscured by the far more newsworthy stories about drunken young women fighting, drinking and 'misbehaving' in public.

Debates about the production of moral panics, and their use, have moved on, however, since Cohen's groundbreaking work and tend to be less conspiratorial in nature. Moreover, moral panics fuelled by the media today fail to hold as much weight. As Kreitzman has suggested, since the late 1960s the 'great and good' have no longer had the power to define the social good, such that morality is now far more socially contested (Kreitzman 1999: 136–7). Indeed, in Thornton's work on youth culture and house music, youth culture is intricately associated with creating moral panics, and to become the subject of one is a sign of being successfully 'anti-establishment'. Thornton suggests that 'with mass media ... affirmative coverage of the culture is the kiss of death, while disapproving coverage can breathe longevity into what would have been the most ephemeral of fads' (Thornton 1995: 122). In a study led by the University of Bath, researchers found that anti-binge drinking campaigns may actually encourage further drunken episodes. Rather than being humiliated or shamed by excessive consumption, bingeing was understood as adding to one's social esteem and was therefore an essential component of forming social identities and networks (ESRC Society Now 2008).

Responsibilization

As noted earlier, one of the problems associated with discussions of binge drinking is that it not only obscures these other activities taking place, preferring instead to focus on extremes of behaviour, there is also little consensus on how the subject is to be framed. As Hayward and Hobbs (2007) have argued, there is a contradiction in the way binge drinking is represented, and in how it is depicted 'as almost the default setting for all young people – including, increasingly, young females' (Hayward and Hobbs 2007: 440–1). Given the complexity of the debate and the multifarious ways that binge drinking functions, it is perhaps not surprising that Jayne *et al.* have called for a deeper understanding of the *context* in which people drink (Jayne *et al.* 2006). Rather than begin from the principle that young consumers are an already-existing problematic demographic, a more nuanced understanding of the actual practices and contexts of alcohol consumption may reveal a great deal more about the role alcohol plays in generating contemporary forms of youth cultures and identities.

The discourses that circulate around bingeing waver between the perception of it as an individual responsibility, or the fault of the state and corporate interests. In government and industry documents, for example, the debate about consuming alcohol has tipped towards attributing responsibility towards the individual. That is, the consumption of alcohol is not the responsibility of the state, but the individual who has freely chosen to pursue his or her own desires. This responsibilization model extends to a deep suspicion about the motives of the state interfering in the home or one's personal life, in terms of a 'nanny state' rhetoric (Redden 2008: 121), and a strong faith in individual agency and free will (Törrönen 2001: 171).

The debate about alcohol and theories of individualization and neoliberalism has become prominent in recent years (Nicholls 2006; Redden 2008; Törrönen 2001). Commonly associated with the work of Ulrich Beck (Beck 1992; see also Brannen and Nilsen 2005; Lewis and Bennett 2003), individualization refers to the contraction of the state in favour of a self-managed and self-actualizing individual. In accordance with broader debates about the meritocracy, this thesis contends that individuals today freely choose their own biographies in life, and this may include both success as well as failure. In this account, bingeing is purely the result of poor self-management and it is not up to the state to control our drinking via punitive measures, licensing restrictions or planning guidance. Rather than being passive 'victims' of an unscrupulous alcohol industry or a government that fails to adequately 'protect' its citizens, the individual is credited with the ability to chose where, when and how much to drink.

This argument is complicated by several factors. In the first instance, inebriation makes the entire notion of free will somewhat less precise (Nicholls

2006). Advocates of the responsibilization model also overlook the ways that consumers are treated very differently by the state. Drinking while young, unemployed and in public is much more heavily regulated and policed and perpetrators face different state and cultural repercussions. The conditions under which people choose to drink are not the same either, and therefore the 'choice' to drink is not a choice that is universally exercised, sanctioned or surveyed in the same manner. The point here is the seemingly obvious observation that drinking – or consumption more generally – is not a value-free practice with the same motivations or outcomes. Who drinks what, where, at what time, under what conditions and in whose company can all fundamentally alter the ways (excessive) alcohol consumption is understood or experienced.

The causes of binge drinking follow a similar pattern, with blame attributed to either poor choices, or to broader structural issues which are seemingly beyond the individual consumer's control.

> Current concerns about 'binge' drinking must be tempered by a consideration of this concerted commercial development and official sanctioning of young adult drinking in the UK over the last decade with the transformation of city centres and the deregulation and elevation of licensed leisure to a new peak in young adult 'time out'.
>
> (Measham 2004: 318)

In other words, the growth of binge drinking is attributed to trends in market forces, the commodification of pleasure and the de-regulation of the drinks industry and licensing restrictions. This approach weighs heavily away from the individual model, and instead attributes causality to aggressive marketers and the normalization of excess consumption. Haywood and Hobbs take a similar approach, but are much more critical of the government for having ignored evidence that questioned the extension of licensing hours (Hayward and Hobbs 2007: 439). In this account, individuals are not free agents able to negotiate the temptations of the late-night city, but are already dominated by powerful corporate forces which are given free reign by absent governments and ineffective legislation.

The reasons behind binge drinking are subsequently understood in competing ways. This is particularly the case when it comes to a potentially addictive substance already laden with cultural baggage about youth, crime and pleasure. In terms of alcohol, then, consumers are either victims of unscrupulous corporate interests, or freely able to negotiate the market and manage their own practices of consumption. To this we could add changing gender relations, where consumption by women has become slightly less pathologized, and different relations of proximity between public and private space, such that drinking in public is becoming more common. Restrictions on access and supply have also shaped the consumption of alcohol, as have other cultural, economic and social trends. Winlow and Hall lean towards the 'depressed and frustrated' explanation (Winlow and Hall 2006), while Haywood and Hobbs claim that consumers are now seeking immediate gratifications. This is reflected in other forms of instant gratification from 'the lottery to *Loaded* magazine' (Hayward and Hobbs 2007: 445). They argue that the night-time economy fills the gap of eternal 'presence' and that consumers drink to escape the over-control and routine of contemporary life (Hayward and Hobbs 2007: 447). Measham and Brain, cited by Hayward and Hobbs, also point towards 'hedonistic cultures' and the resulting 'normalization of recreational drug use' (Hayward and Hobbs 2007: 441; see also Measham and Brain 2005: 266).

Needing to 'blow off steam' has a commonsensical ring to it, but it does not explain why some people chose to indulge – or relax – in this manner. As much as drinking may be an enjoyable way for some people to unwind, some individuals may choose other activities. Others may find the taste of alcohol or the sensation of intoxication far from relaxing, despite having a similar need to unwind. The focus on hedonistic cultures or liminal and transgressive behaviours (Hayward and Hobbs 2007: 444) also does not account for how these hedonistic cultures emerged in the first place or what was specific to the late 1980s and early 1990s that gave rise to the normalization of hedonistic cultures. Planning, legislative changes, post-industrialization and marketing practices are figured to be contributing factors, but again this does not entirely explain why some people who lived through the culture of hedonism, or the 'summer of love' where drug taking became normal *for some*, rejected or embraced those cultures.

Marketing

Advertising, especially that which is targeted towards young people, is a key argument here. However, as the *Safe. Sensible. Social.* report states, citing research conducted by MORI for the Portman Group, while seven out of ten people feel advertising of alcohol affects the amount other people drink, only one in ten believe it effects them personally (HM Government 2007: 14). This disjuncture demonstrates in stark terms the way that consumers are infantilized by claims that advertising has a direct influence on consumption. Young drinkers are rendered as both immature in their desires and wants, and as lacking any form of agency in negotiating the marketing of the brewers. It furthermore removes advertising and alcohol consumption from the broader culture in which both operate, ignoring how the media functions within a matrix of relations pertaining to work, choice of peers, family life, gender, age, community and access.

At the very least, advertising is just one component of promoting alcohol, and other aspects of the marketing process, such as at the point of sale, in-store promotions or free giveaways should also be examined. The focus on the drinks industry and advertising as responsible for binge drinking should not detract from other inducements to drink to excess. Peer pressure, the practice of buying rounds and different manifestations of masculinity were all found by Lindsay to encourage excessive drinking in Melbourne pubs (Lindsay 2005: 41). It was other consumers, in other words, who were also actively promoting binge-drinking episodes.

The belief that young people are easy prey for alcohol producers and advertisers returns us to rudimentary media theory and the notion that consumers, especially young women, are especially susceptible to media imagery. The 'hypodermic syringe' model has now been generally discounted by media theorists (Gauntlett 1998) but it reappears with regularity once the debate turns to young people's drinking patterns. The problems with this passive model of audience reception are various and have been extensively documented for several decades. Of key importance, it does not explain why some consumers are more susceptible to advertising than others.

Gunter *et al.* note that there is 'mixed evidence' as to the role of advertising in alcohol consumption (Gunter *et al.* 2008: 8). Given this current impasse, a more fruitful line of enquiry could proceed from the basis that advertising is but one aspect of the entire alcohol narrative and experience. For advocates of the 'advertising promotes drinking' model, research is clearly needed that examines the extent to which advertising plays a role in shaping consumer desires, which consumers, what sort of drinks and in which contexts. That is, to what extent does advertising play a role, compared to, say, bar design, class, gender, race, geography, the company in which consumption takes place or indeed the actual taste (Gunter *et al.* 2008)?

Pricing and consumption

> The fact of the matter is price is a crucial determinant of how much we drink.
>
> (Don Shenker, Alcohol Concern)

Pricing is one of the more common methods advocated for minimizing alcohol misuse. Essentially, the issue of pricing comes down to taxation which, as Plant and Plant note, is second only to outright prohibition in terms of the controversy it generates (Plant and Plant 2006: 142). The logic behind increasing taxation is quite simple: the more expensive the product, the less likely people are to buy it and alcohol has proven to be sensitive to both price reductions and increases (Plant and Plant 2006: 142).

Alcohol Concern, in their 2008 Budget Submission, called for a blanket 10 per cent tax increase on all alcoholic beverages. The evidence for this call comes from the Academy of Medical Sciences' *Calling Time*, which claims that 'price moderation, usually through tax increases, is highly effective, particularly in under-age drinkers' (Academy of Medical Sciences 2004: 7). The report also recommends 'increasing tax on alcoholic beverages to disposable income' (2004: 9). The AMS claim that a rise in alcohol prices by 10 per cent would result in a drop in cirrhosis of the liver to the tune of 7 per cent in men and 5 per cent for women, and that deaths by alcohol-related causes would drop by 38.8 per cent for men and 37.4 per cent for women (2004: 27). While there is no way of guaranteeing this outcome, on the basis of these figures it would appear to be a highly effective measure. It has also become increasingly popular. Under the 2008 budget, Chancellor Alistair Darling increased alcohol excise duty by 6 per cent against inflation, with a further 2 per cent above inflation after 2008 (*Financial Times*: 12 March 2008).

The AMS concedes that taxation imposed on alcohol impacts on heavy drinkers and the under-age more than everyday drinkers – a group which, in any case, are poorly understood. Plants and Plant, drawing upon a number of international studies, also suggest that it is the under-age and heavy drinkers who are most likely to respond to increased prices. However, an increase on taxation of alcopops by the Australian government (Standing Committee on Community Affairs 2008) has not been without controversy. Concern about the consumption of alcopops in Australia was given credence when the Australian Division of General Practice argued that 'forty-five per cent of girls as young as 12 ... said their last drink was an alcopop, compared to eight per cent saying it was beer and 11 per cent wine' (Australian General Practice Network 2008). The subsequent move to impose a 70 per cent tax on alcopops, while popular in some quarters, was questioned by other groups. One particular concern was that consumers would simply choose to buy their own spirits and mixers, and thus 'free pour' their own alcopop-style drinks. The liquor industry in Australia claim this is precisely what has taken place. The sale of alcopops has decreased by 30 per cent since the tax came into affect in 2008. The sale of spirits, however, has increased by 46 per cent (Munro 2008). The Australian Medical Association have rebutted any link between the two figures. A further problem is determining how the industry will respond. Already in Australia, the 70 per cent tax on alcopops has allegedly been met by some manufacturers with a reduction of alcopops based on sprits and the introduction of alcopops based on wine. The wine industry in Australia receives subsidies and tax breaks, meaning the 70 per cent can be avoided. It is yet to be determined how successful wine-based alcopops are, but Godfrey has noted that blanket increases in taxation do affect different alcoholic drinks in different ways. Beer, for example, is 'price elastic', as is, increasingly, wine (Plant and Plant 2006: 143).

Were taxation to be increased, supermarkets would have to oblige, but there is no guarantee they would not continue to provide a more economical alternative to bars and clubs. As the director of a large nightclub chain noted:

> The trouble is, if they've got a minimum price on it, it then becomes 20p more expensive to sell a pint of lager. People won't go out – they'll sit at home and go to the off-license, buy 6 cans of lager [to drink] before they go out, which is what's happening now.
>
> (Interview 2006)

Young people on limited budgets who wish to drink may continue to patronize outlets that offer them the most economical option. And, there is an argument to be made that drinking in licensed venues under supervision and community control (Valentine *et al.* 2008) is preferable to drinking in parks or on the street.

In 2007, the three main supermarket chains in Britain, Tesco, ASDA and Sainsbury's were heavily criticized for their pricing of alcohol. The three were selling beer for the equivalent of 50p per litre. This was less than mineral water, which sold for 56p per litre, or soft drinks such as cola also at 56p per litre. The story, as it eventuated, was actually slightly overblown in that the water in question was a high-quality brand while the beer was a 2–3 per cent 'home brand'. Nonetheless, the supermarkets concerned were accused of selling below cost and admonished by government and industry groups.

While not wishing to advocate the selling of cheap liquor, supermarkets are in a difficult position. Rather than increasing taxation, local councils or indeed sellers could work to increase general prices, but, to date, this has failed. When the council of Aberdeen, Scotland attempted to impose a minimum pricing scheme in 2004 they were taken to the Court of Session by the Spirit Group and Mitchells and Butler, both being large pub and bar companies. The council had attempted to impose a minimum price of £1.75 on a pint of beer, lager, cider or alcopops, and £1.20 for spirits. Their action, however, was deemed 'unlawful' (Anonymous 2004a). Supermarkets could similarly find themselves taken to the Office of Fair Trading if they were to 'artificially' increase prices. That is, if large outlets worked together to set a minimum price for alcohol, they run the risk of being charged under unfair competition laws (BERR 2008). In early 2008 the supermarket chain Morrisons did call on the government to examine competition laws in relation to the pricing of alcohol. It is yet to be determined how successful this has been.

The most spirited argument against increasing pricing comes from within the alcohol industry. According to the Wine and Spirit Trade Association, Britain already has the highest level of taxation on beer, the second-highest level on wine and the third-highest level on spirits. In Italy and Spain, often held up as ideal drinking cultures, there is no tax at all on wine, and in France tax accounts for 20p per bottle.

Summary

In the UK, there is an often contradictory tension around alcohol whereby it is both a structural problem attributed to poor government or 'alcoholic genes', and, increasingly, to the fault of the individual to adequately self-police their alcohol intake. Alcohol has, in these terms, become a matter of self-regulation rather than explicit management. AHRSE, for example, does not seek to reduce the national alcohol consumption figures, but rather to tackle misuse and harm. It assumes that the problem of excessive drinking already exists and it is the role of government to deal with the consequences, rather than take a proactive stance before the problems begin to emerge (Heather 2006: 230; see also Room 2004). The government has sought, therefore, to manage the problem by striking the right balance between 'advice' in the form of labelling on the one hand, but liberalizing the industry on the other and thus passing full responsibility for over-indulgence into the hands of the consumer.

This chapter has examined the various problems in defining binge drinking and the often contradictory and competing ways it is understood. Rather than posit a source of the problem – advertising, boredom, genetics or pricing, for example – we have instead sought to examine the complexity of the debate in the hope of raising further questions. One final question concerns the overall debate, as it stands, on binge drinking in Britain today. Binge drinking may well be common in some segments of the community, but the majority of people do not binge drink. One particular problem with the continual focus on extremes of behaviour is that it makes it difficult to clarify an even larger question concerning the boundaries between bingeing, normal drinking, hazardous or harmful drinking. As Drummond argues, the AHRSE suggests that alcohol misuse is a 'minority problem', and that strategies such as targeted policing are the solution: 'Alcohol problems exist across the age range and are not restricted to people frequenting city centre pubs on Friday and Saturday nights as the Prime Minister's analysis of the problem implies' (Drummond 2004: 378).

In policy and mainstream media terms, reports on binge drinking have eclipsed discussion of everyday patterns of alcohol consumption, making it exceedingly difficult to actually define and explore 'ordinary' drinking, especially amongst older or middle-class consumers. As Measham has observed:

> In policy and practice terms, consumption needs to be conceptualized as a spectrum of possible positive and negative consequences for the user, for their associates and for wider society, rather than as

> a problematic–unproblematic dichotomy implied by terms such as 'binge' drinking which serve to emphasize the credibility gap between the realities of young people's leisure time consumption and the health education advice directed at them.
>
> (2004: 316–17)

Binge drinking is an emotive issue, an often imprecise term and, as deployed by some media reports, exceedingly divisive. It is overburdened with anxieties about late-night disorder, youth, and changing social, economic and gendered relations. We return to more everyday experiences of drinking in Chapter 9. The following chapters develop the theme of regulation, turning to measures designed to better manage and regulate the consequences of the night-time city.

Chapter 6

Regulating consumption

Mainland Britain

[Press] Question: Some people are worried you are a bit of a moralist/puritan killjoy.... Can you give a guarantee that you won't mess with our 24-hour drinking laws or do you think they are in fact leading to increasing violence and ill health and have got to be reviewed?

Prime Minister: I have been in London for more than 20 years, and I am reminded of the story that was told about Mark Twain when he went to Nevada and he arrived from a very puritan background, very church-going family, he arrived in Nevada and he found drinking, and gambling and womanising and everything else, and he said 'This was no place for a puritan, and I did not long remain one.'

(10 Downing Street 2007)

Introduction

The period 2001–2005 saw changes to licensing legislation in England and Wales. These changes were bureaucratic in origin, but were subsequently 'spun' by government as presaging a more profound cultural shift. Although the passage of the legislation through Parliament attracted little national attention, the furore that was eventually aroused, after the legislation had reached the statute book but before it was implemented, laid bare key controversies over public drunkenness and the extent to which localities can control public comportment and private interests.

Hadfield's (2006) excellent account of the passage of the Act into legislation sets out the contestation that took place over licensing legislation, charting the government's move towards neo-liberal de-regulation and its subsequent accommodation of objections. This chapter will consider the aspects of the Act that have greatest relevance to the regulation of the built environment in addition to re-telling such elements of the story as are essential to an overall understanding of its impact on the urban landscape. Changes in licensing legislation brought forward some interrelated concepts and regulatory practices that lie at the heart of micro-management of the evening and night-time economy. Spatial and temporal controls are present in licensing legislation and are intertwined with regard to their impact and effects. This chapter will scrutinize the evidential base from which judgements were made and are still being reviewed.

The Licensing Act (2003) fundamentally changed the processes through which applications for liquor licences are considered and determined. This change to the extent to which local control is permitted and feasible lies at the heart of debates within planning and governance. Assessments of the impacts of the Act to date will be reviewed. The changes to licensing legislation in Scotland provides a further reminder of the importance of local devolution and an illustration of how the English and Welsh legislation might be improved.

Changes in licensing legislation: an overview

The origins of recent changes in British licensing legislation derive from a convergence of private interests and an overriding theme of government desire to cut down on bureaucracy and 'red tape'. The 1964 Licensing Act had set prescribed hours for the consumption of alcohol in licensed premises – that is in pubs, clubs, bars, restaurants and cafés – and for the sale of alcohol to be consumed off-premises. Opening hours remained restricted and premises were compelled to continue to close in the afternoon. Afternoon closures had been in effect since the First World War; a result of concern about inebriated ammunitions and shipbuilding workers (Barr 1998: 140). Review and liberalization of the licensing laws was considered in 1972 by a task force whose report recommended many changes that were subsequently implemented in a piecemeal manner. The most dramatic of these was the removal of the concept of 'need', which, as discussed in Chapter 3, was eventually removed in 1999. Opposition from medical professionals and public health pressure groups blocked full implementation of the report (Light 2005).

The Licensing Act of 1988 permitted continuous opening hours for weekdays, thereby allowing the opening of continental-style cafés between the hours of 11 a.m. and 11 p.m. in London. This liberalization of hours was extended by a revision of the legislation that permitted all-day Sunday opening via the Licensing (Sunday Hours) Act 1995. The election of a Labour government in 1997 saw the furtherance of an idea to streamline legislation and do away with unnecessary regulation. In 1998, the Better Regulation Task Force (1998) undertook a review of licensing legislation. The review took the view that the main objectives of licensing legislation were to prevent the public from noise and disorder and to prevent citizens, in particular young people and children, from harming themselves. The review furthermore recommended primary legislation that would simplify the law. At the time of the 1998 review, licensing legislation was complex and labyrinthine. This posed problems for independent licensees, who found that the operation of a small nightclub could require the acquisition of up to eight types of permission for liquor and entertainment licensing alone.

The sale of alcoholic refreshment has been regulated since the end of the fifteenth century. Until the Beer Acts of the 1830s, justices of the peace (magistrates) had powers over the numbers, locations and hours of operation of licensed premises (Talbot 2007). In 1998, control over the sale of alcohol was bifurcated. Magistrates had powers to determine whether an applicant could be granted a licence to sell alcohol, either for consumption on the premises (an 'on-license') or for consumption away from the premises (an 'off-license'). The hours at which alcohol might be sold were regulated by central government, with different regimes operating in England and Wales, Scotland and Northern Ireland. In England and Wales, local authorities could permit nightclubs and other premises to open beyond the prescribed closing time, the 'terminal hour'. For venues to open later, they needed two licences, a public entertainment licence, which permitted music and dancing, and a special hours certificate, which permitted the sale and consumption of alcohol. An anomaly of this arrangement was that the sale of food was required to accompany the later sale of alcohol, although in practice, as Hadfield recounts, the requirement was frequently circumvented through wily licensees offering one (inedible) dish (Hadfield 2006).

It was not only licensees that faced problems with the operation of licensing in this period. Objectors faced many difficulties in taking action against powerful and wealthy companies (Central Office for Information 1998). Since the only redress under the law for continued infringements of a licence was revocation, it could mean removing someone's livelihood, which was a step only to be taken as a last resort.

The 1998 Better Regulation Task Force report recommended primary legislation because they argued that greater transparency in decision-making would be achieved by taking licensing powers away from magistrates and transferring the responsibility to local authorities. The licensing system would also become more accessible for licensees and objectors. On the issue of hours, the committee was more circumspect and recommended greater flexibility for later and earlier closing, but this was to be decided based on criteria that 'focus[ed] on the need to prevent nuisance and disorder and to protect young people' (Central Office for Information 1998: 3).

The first draft of primary legislation followed soon after in 2000, in the form of a White Paper entitled *Time for Reform: Proposals for the Modernisation of our Licensing Laws* (Home Office 2001). The Home Office produced the paper. At this time, the Home Office was a large ministry responsible for policing, prisons, immigration and the entire criminal justice system. As the draft legislation became law and controversy developed, different parts of the governmental machine were drawn in. The Home Secretary's foreword to the paper made an acknowledgement of other interests. After succinctly summing up the underlying tensions between allowing greater flexibility on the sale of alcohol, reducing crime and disorder and protecting residents and children, the foreword noted that, 'the decisions we make on these issues will in turn help to shape the future of our villages, towns and cities' (Home Office 2001).

The White Paper adopted the recommendation that powers for licensing be unified and handed over to local authorities. A simplified system of personal licences for licensees and premises licences for different types of location where alcohol might be sold would also be established. The most radical departure of the proposed 'modernization' was that the system of opening hours would be entirely abandoned and each venue, whether a bar, pub, club or supermarket would be allowed to apply for its own operating hours.

The White Paper was put out for consultation in 2000, before being re-drafted as a Bill to be taken before Parliament. Responsibility for the Licensing Bill was taken over by a different ministry, the Department of Culture, Media and Sport. This transfer of responsibility was not simply an administrative convenience, but signalled a desire on the part of the government to position licensing controls as a component of tourism and economic development. The benign aspects of alcohol consumption as part of a vibrant public culture were therefore given emphasis, with the underlying message that licensing was no longer associated with heavy-handed control, but was to be considered instead a part of everyday 'cultural' life in a sophisticated society.

As a senior civil servant explained:

> The first thing I should say is the machinery of government changes are not discussed with civil servants at all except I think in the Cabinet.... I can't say I am a great authority on this – I believe the political thinking would have been that it needed a more balanced

approach when it went into parliament so that all the issues were addressed whereas the Home Office tends to be forced into a corner to talk about crime and nothing else. That was a big part of the Act, things like entertainment, theatres, concert halls, licensing and public space. It doesn't make newspaper headlines but it has been a really important part of what we do.

(Deputy Director – Industry (Tourism, Economic Impact and Licensing) DCMS)

In doing so, greater influence was accorded to the licensed trade, as will be discussed further in this chapter. The Licensing Bill adopted the main provisions of the White Paper and set out a new framework for liquor licensing.

The framework set out within the law was set to operate within four major objectives. These are:

- To prevent crime and disorder.
- To ensure public safety.
- To protect children from harm.
- To prevent public nuisance.

Local authorities were also required to set out statements of licensing policy. These statements were a continuation of the type of statement that had been previously issued by magistrates in each area. Local authorities had to consult widely on these statements, but the essence of each was a statement of licensing objectives, which were statutory, and the means by which they would be achieved.

The Licensing Act was accompanied by national guidance to local authorities. Parliamentarians demanded that a draft of the guidance accompany the Act before it passed on to the statute book. The *Guidance* (Secretary of State for Culture 2002) was amended as the Bill proceeded through Parliament and was amended further before the Bill was implemented some 18 months later and has since had two subsequent amendments. The number of amendments is a reflection of the controversies that surrounded the legislation following its passing.

Public debate about the Licensing Act was relatively muted during the passage of the Bill through Parliament in 2002–2003. A Parliamentary Inquiry was established by the Office of the Deputy Prime Minister (the equivalent of a Ministry for the Environment) and this heard evidence from residents' groups, the licensed trade, the police, local authority leaders and alcohol campaigners. Its final session allowed committee members to interrogate four government ministers from the ministries most relevant to the built environment, Transport, Culture and Sport, Housing and Planning and the Home Office (policing). The Inquiry was specifically set up to probe the interaction of an expanding night-time economy with the policy to introduce more residential uses back into town and city centres (House of Commons 2003a). The Committee made 34 recommendations in its final report, some of which, such as changing the planning use class orders, were implemented, and others, such as banning 'happy hours', were not. Ironically, it was only after the Act became law in July 2003, but before it was due to be fully implemented in November 2005, that media awareness heightened. At that point, apart from delaying the implementation of the Act, there was little that could be changed, apart from the *Guidance*.

The licensing act and the media

The aspect of the legislation that preoccupied the media was the removal of permitted hours. This they dubbed '24-hour drinking', a catch-phrase that came to dominate TV interviews with ministers as well as newspaper reports. One of the earliest outputs from the media had in fact come from an academic. In a catchily headlined article, 'Mayhem After Midnight' published in a Sunday newspaper in 2002 Hobbs (2002), he gave an accurate description of the type of incivility and disorder prevalent in an unnamed British city. The spectacle of youths vomiting in municipal flowerbeds, urinating and fighting in the early hours of the morning was then unknown to the average *Sunday Times* reader, who might expect to be in bed by midnight.

The year between the appearance of Hobbs' article and the passing of the Licensing Act saw increased activity on the part of lobby groups, police representatives, alcohol campaigners, residents' groups and charities. Reports of the issues surrounding the Act tended to the superficial, however. The government's viewpoint was reported uncritically, drawing on government press releases. It was only in isolated comment pieces that rare voices of dissent appeared (Kettle 2003). In 2004, an article appeared about a 'leaked' letter from the then Home Secretary to the Prime Minister warning of town and city centres spiralling out of control through increased drunkenness (BBC News 2004).

Throughout the remainder of the year, more articles appeared in the press, while television programmes documented levels of drunkenness, crime and disorder prevalent in many town and city centres (see, for example, Levy and Scott-Clark 2004). This reached a crescendo early the following year when more than half of the regional police forces publicly expressed their criticism of the Act and concerns for its consequences (Taylor and Hughes 2005). The leading article in the *Daily Mail* thundered 'This ill-judged law MUST be stopped' (13 January 2005: 2) and devoted seven pages, including the front page, to the case for opposition. Seven police chiefs, five city councils, distinguished physicians and the leaders of the two opposition parties supported its 'campaign'. ACPO, the Association of Chief Police Officers and the charitable group Alcohol Concern, similarly questioned the logic behind the legislation. Former Conservative party leader Michael Howard went as far as proposing the Act be held back until concerns about alcohol consumption had been adequately tackled.

In response, the government as a whole shifted its ground with regard to the regulation of alcohol. It appeared that there was debate going on behind the scenes. A newspaper article commented on the 'war between ministries' and it would appear that the Home Office was clashing with the Department for Culture, Media and Sport over licensing (Robbins 2005; White and Hetherington 2005). Immediately following the media 'campaign' a joint report, *Drinking Responsibly* (Department for Culture, Media and Sport *et al.* 2005) was issued by three ministries (loosely, Home Office, Culture and Environment) and subsequent government pronouncements attempted to demonstrate 'joined-up thinking'. The *Guidance* to the Act was amended and a strengthened discourse introduced on the part of ministers that cautioned against excess, and recognized alcohol-related disorder. Government statements emphasized police powers within the Act to close down disorderly premises and powers awarded under other legislation to combat anti-social behaviour, such as dispersal orders, fixed penalty fines and banning orders.

Still, the press comments intensified throughout 2004 and prior to the full implementation of the Act in November 2005. The contradictions in the government's attitude towards alcohol were summed up by a journalist who neatly exposed the gaps between aspiration and reality:

> On the one hand, Ministers encourage ever more access to booze by extending pub hours, while on the other they lecture us on the perils of binge drinking. Jack Straw, when he was at the Home Office, said long pub hours were meant to cater for 'men who could take their drink', while Tessa Jowell [Minister for Culture] posed for the press pulling pints. Yet they fret and fume that our town centres are turning into war zones, impossible to control or stop.
>
> (Alibhai-Brown 2005: 15)

The media discussion focused on crime, disorder and the intersection of private interests and civil society. It was unable or unwilling to tackle the more complex aspects of regulation with regard to the built environment and the necessity of regulating not one, but two, intersecting dimensions, space and time. To look at these in more detail, it is necessary to return to the evidence that the government used as a basis for its regulation.

Problems with the evidence base

Lobby groups and academics introduced important aspects of the debate in environmental terms. These revolved around space, the number of licensed premises, where they are situated, in what concentration and their proximity to different land-uses. The Licensing Act presupposed that the spatial aspects of licensing would be dealt with through the planning system. Indeed, the Act has the explicit intention of not duplicating other legislation. The planning system in England and Wales is not retrospective, however, and hence can only deal with new applications. Pre-existing spatial considerations are therefore inextricable from temporal controls.

The regulation of licensed premises is not only about protecting citizens from harm to themselves, but also about the prevention of harm to each other through crime, disorder, anti-social behaviour and nuisance. Tiesdell and Slater (2006) draw on the concepts of 'dispositional and situational approaches' to crime reduction in their consideration of crime reduction through licensing and planning regulation. While planning legislation cannot generally influence an individual's disposition or motivation towards crime, it makes a direct intervention into the circumstances or 'situation' in which a crime might be committed. Tiesdell and Slater point out that licensing regulation is an important component of a situational approach to crime reduction and, by changing the environment in which people drink, behaviours can be modified. While town planning historically deals with the spatial, the authors highlight the importance of the temporal dimension and, in particular, the differences between the evening (5 p.m.–9 p.m.), the night (9 p.m.–midnight) and the late night (post-midnight). In a similar vein, Graham and Homel (2008) consider a situational approach. They draw on studies that highlight two different circumstances in which the possibility for violence is increased. These are, first, 'clustering points', where people congregate and remain for some time, for example in queuing for taxis or late-night buses, or for fast-food takeaways. The second category is 'congestion' points, where people are moving in large numbers in overcrowded circumstances, such as exiting from a large nightclub through a narrow street, or where several licensed premises close simultaneously. These concepts are important in understanding the limitations of the Licensing Act, 2003.

The recommendations of the Act appear to have been based on two research papers. Deehan (1999) summarized evidence from studies in the 1990s in a Home Office research paper. Drawing on Australian and British research, the report noted that studies carried out in the 1980s and 1990s had found an association between violent crime and licensed premises. The manner in which licensed premises were run had an influence on levels of violence and disorder, such as swearing and noise both inside the venue and immediately

6.1
**Crowds outside a
nightclub in
London on a
Friday night.**
Source: photograph
by author.

adjacent to it. Where venues were overcrowded and smoky, with poor access to the bar, levels of violence were exacerbated. Irresponsible practices, such as price discounting and serving already-inebriated people were also factors, as was the behaviour of door staff and bar tenders and the levels of anti-social behaviour that they were prepared to condone. The problems of aggressive door staff were also acknowledged. It was noted that these problems applied to a small number of premises, mainly located in the micro-entertainment districts within town centres. The recommendations of the review therefore focused on ensuring that all licensed premises were run to an acceptable standard.

In the lead up to the passing of the Licensing Act, the problems of concentration were only considered with reference to Marsh and Kibby's (1992) research. Their study, which was carried out for the Portman Group, the body representing the licensed trade, provided an in-depth view of the problems of crime and disorder associated with drunkenness. The research was qualitative and based on five English towns and small cities in which alcohol-related disorder had been reported. Two seaside towns were also investigated, to a lesser degree. The report finished with a cross-European comparison drawing on further studies in Holland, Italy, Spain and France. The first recommendation of the report was of the importance of changing closing times. The rationale was straightforward. Marsh and Kibby noted that over 50 per cent of all arrests for alcohol-related crimes and Public Order Offences occurred an hour after the 11 p.m. terminal hour on Friday and Saturday nights (Marsh and Kibby 1992: 155). A further peak of violent disorder occurred just after nightclubs closed.

These peaks were due to the densities of drunken people on the streets. This had five consequences. First, the potential for violence was increased, and second, there were opportunities for individuals to 'show off' to large, encouraging audiences of bystanders. Third, the police had more difficulties in dealing with offences and a small disturbance had the potential to turn into a major incident. Fourth and fifth, competition for scarce resources, such as transport and fast-food, led to more conflict. To exacerbate matters, taxi-drivers were unwilling to operate at this time of night.

Drawing on experience from the Netherlands, where experiments in 'free' or extended hours had been tried, the authors recommended that: 'On the basis of this evidence, we propose that amendments should be made to the relevant Licensing Acts to permit experimental trials of either extended or deregulated licensing hours in certain local areas' (Marsh and Kibby 1992: 157).

These experiments were to be carried out only in areas where licensees agreed to operate responsibly, with high standards of management and a clampdown on under-age drinking. They also made it clear that they would expect such experiments to be carried out in areas where licensed premises were not in close proximity to housing.

Marsh and Kibby anticipated the results of this experiment to be a reduction in alcohol-related violence due to a lowering of the densities of drunken people on the street at any one time, and a reduction in drunkenness since there would no longer be a requirement to 'drink up' before closing time. The authors also anticipated that customers would gradually 'drift away' from premises towards home, mainly in a relatively peaceful manner. Other benefits would also accrue, for example, demand for transport and fast-food would be more evenly spread and therefore easier to provide.

In carrying forward her account of Marsh and Kibby's work, Deehan (1999) highlighted the recommendation for flexible drinking hours. The rationale that she provided was phrased in terms of avoiding the 'peaks' of violence. However, she did not make explicit the link between densities of people on the street and the potential for violence, disorder and anti-social behaviour, leaving it to be implied. This elision was to become important in arguments about the Act and in its aftermath.

Two further Home Office papers were published while the Licensing Act made its way to the statute book. A review of statistics gathered in a national survey, the British Crime Survey, between 1996–2000 was presented with regard to alcohol-related assault (Budd 2003). The British Crime Survey is based on interview data and incidents not reported to the police. While making the point that a very small proportion of adults, 2 per cent in 1999 (Budd 2003: iv), were the victims of such assaults and that the number had dropped between 1995 and 1999, such offences were 'by no means trivial' (Budd 2003: v) and many happened in the context of the night-time economy. Typically, young men were involved as victims and perpetrators. The report concluded by arguing that the situational measures conducive to violence should be tackled as a policy measure. A further review, based on two separate studies (Richardson *et al.* 2003) confirmed the desire amongst young people to binge drink, and the result of this behaviour as being offending and disorder. One of the studies, a qualitative piece of research using focus groups, highlighted the factors that led to being drunk and disorderly. These included temporal aspects, such as lack of late-night venues, and spatial influences, including 'poor town centre layouts' and lack of late-night transport. Neither of these research reviews directly supported removal of permitted hours.

The Metropolitan Police (MPS) in London also raised concerns about the prospect of the Licensing Act. An internal report argued that closing times were not the only causal factor in binge drinking (Metropolitan Police Clubs and Vice Operational Command Unit 2003). The report made particular reference to a study carried out in Cardiff in 2000. Entitled 'Tackling alcohol related street crime' (TASC) the project combined data from police sources and the local hospital. The study found a pattern of violence whereby alcohol-related incidents were heavily concentrated on Friday and Saturday nights. In this, it followed Marsh and Kibby's study. Where the TASC study differed was in the times at which offences occurred. Rather than being concentrated between 11 p.m. and midnight and 2 a.m. and 3 a.m., they stretched through the night from 11 p.m. to 5 a.m.

The MPS commented that premises were likely to still close at a specified time. The gradual drift towards later hours in central London had stretched police resources, such that the commander of the central area told a conference that at any one time, he might only have seven officers available to deal with crowds in the West End in the early hours of the morning. Although this situation had been gradually improving, with a shift of resources towards the early hours, an extension of operating hours later into the night could stretch policing further in other boroughs throughout London and take resources away from other types of crime prevention.

The Licensing Act 2003 took up Marsh and Kibby's proposal for de-regulating hours, ignoring their suggestions for an experimental period and series of trials. As McNeill points out, the change to completely unrestricted hours was unprecedented in the rest of the developed world. Others commented that Marsh and Kibby's research had been carried out before the alcohol and entertainment industries had re-structured in the late 1990s. That is, the youth-orientated alcohol market was not fully established at the time Marsh and Kibby were conducting their research. Aggressive marketing tactics, such as the selling of 'shots' and alcopops had not been developed, nor was there a comparable number of youth-focused bars and clubs in town centres. The British Entertainment and Dance Association (BEDA) commented that the loosening of regulations preceding the introduction of the Act had encouraged a proliferation of premises. They also argued that competition between operators had resulted in price discounting, which exacerbated binge drinking. This price discounting occurred earlier in the evening, during 'happy hours' (Institute of Alcohol Studies July 2007). One further problem was that Marsh and Kibby's study had been carried out in smaller towns and cities, such as Oxford and Preston, before the advent of large-scale premises where three or four major nightclubs might be located in a relatively small area.

Fhe police raised the issue of 'transient' drinking, by which they meant revellers 'circuiting' between pubs, bars and clubs (Metropolitan Police Clubs and Vice Operational Command Unit 2003). This practice, which is part of a normal night out for many people, means that there are many people in varying rates of intoxication on the streets at any one time, moving between different venues. Later closing hours would not reduce numbers or necessarily the practice of circuiting; rather it was likely to extend the times at which it occurred. That the government was aware of this problem was demonstrated by comments made in the first and second drafts of the *Guidance* issued as an accompaniment to the Licensing Act. This stated that 'fixed' and 'artificially early' closing times were to be avoided as it was argued that customers would transit from one zone to another, thereby causing disturbance to residents (Secretary of State for Culture 2002). Evidence was brought forward from the experience of extending licensing hours in Edinburgh to support this. 'Engineered' staggering of hours, by allocating terminal hours of say 1 a.m., 2 a.m., etc., to particular premises was also not recommended because it would create smaller peaks of dispersal. Instead, a gradual 'natural' dispersal through longer hours was to be encouraged.

Campaigners argued that notions of 'natural dispersal' in the context of a market-driven night-time economy were misplaced. In a commercial context where major pubcos were running large warehouse-style drinking 'barns', it would not be feasible for them to keep a place open with only a few customers. Managers would not be willing to keep staff on, waiting for the last customer to leave (Institute of Alcohol Studies July 2007). While such a proposition might just be possible for a small owner-managed bar or restaurant, it was not a strategy that made any sense for a high-volume venue that employed a significant team of staff.

The other flaw in the notion of a quiet 'natural' dispersal was that of sheer numbers. The expansion of numbers of licensed premises in particular centres had, as previously noted, led to high levels of pedestrian density. When there are as many as 80,000 people on the streets at once, even one-tenth of that number might still be present in the early hours of the morning, sufficient to create a disturbance even when simply walking around without shouting, swearing or vomiting. The situation is compounded when that number of people is concentrated in a small space in 'congestion points'. On this, the government made two important concessions.

Cumulative impact and use class orders

'Cumulative impact' is a term coined from planning guidance (Department of the Environment 1996) and refers to the negative impacts that an agglomeration of licensed premises might have. The impacts might be noise, increased litter from fast-food takeaways or increases in disorder as more people encounter each other. The White Paper and draft guidance did not recognize this condition at all, and instead focused on the individual licensed establishment. The supposition was that planning legislation would prohibit concentration. In fact, the White Paper argued that licensing authorities should not re-run planning applications. Applications for licences should be determined solely on their impact on crime, disorder and nuisance and 'commercial matters such as economic demand' should not intrude on the decision-making process. This view simply did not recognize the problems associated with existing areas of 'saturation'.

These were succinctly summarized by Westminster City Council in its evidence to the Select Committee. They argued that planning authorities had few opportunities to contain or control growth in alcohol-related entertainment. The possibilities for 'migration' between different types of entertainment use were too great. Statutory arrangements were too long-winded and inflexible to respond to the dynamism of the late-night economy. For example, where new planning permission had to be sought, the application would be determined based on the policies in the Unitary Development Plan. If saturation had occurred following the adoption of that plan, the policy could be changed only by attending a public inquiry. A wily operator could also breach planning conditions and by 'playing the system' could extend the enforcement process to 18 months. The submission concluded 'planning legislation currently provides little protection for authorities seeking to manage the late-night economy' (House of Commons 2003b).

Many objections were made to the National Guidance Sub-Group of the Licensing Bill Advisory Group, and a debate was held in the House of Lords. This gave rise to a somewhat 'grudging accommodation' in a revised version of the *Guidance* to the Act (Hadfield 2006). The new clauses allowed local authorities to declare a limited area in their jurisdiction as a 'special policy' area, providing that there was statistical evidence that a concentration of licensed premises was having an adverse effect on crime and disorder (Secretary of State for the Department for Culture, Media and Sport 2004). The designation would be submitted as a component of the authority's licensing policy.

As was discussed in Chapter 4, one of the reasons that local planning authorities had overseen a situation where concentrations of licensed premises had emerged was the looseness in the statutory regulations controlling land-use. The UK planning acts controlled changes to land-use through a system of categorization. This was not set out in primary legislation, but in statutory guidance. Planning permission had to be sought for new developments and for certain types of conversions, extensions and changes between categories. Unlike most of continental Europe, in the UK system planning control is separated from building control and other types of regulation, such as liquor licensing. The 'migration' between different types of use that was permissible under planning control meant that a cinema could change into a nightclub, subsequently obtain a public entertainment licence and a special hours certificate and open until the early hours of the morning. Restaurants could also change to café bars and ultimately dance bars without the need to apply for a change in use. The impact of these changes on a mixed-use neighbourhood was considerable.

The government sought to close this loophole by changing the Use Classes Order (UCO). A junior minister announced in March 2003, while the Licensing Bill was being debated, that revisions would be made that would prohibit the conversion of restaurants into bars, placing drinking establishments into a different category. Following consultation, hot-food takeaways were put into a separate category, apart from restaurants and cafés. This change was important as the Home Office studies cited earlier had found that queues for fast-food takeaways were frequently flashpoints for violence. Nightclubs were placed as *sui-generis*, that is, in a special category of their own, such that planning permission is required for any change of use either to them or from them. The changes were not fully enacted until April 2005 when they included other changes relating to offices and retailing uses. Ministers explained that the changes concerning the night-time economy would enable local authorities to 'get the right balance of businesses on the high street' (ODPM 2005a); on the one hand 'boosting' the evening economy and on the other controlling a proliferation of bars and fast-food takeaways.

In summary, the passage of the Licensing Act saw a loosening of temporal controls, which applied to all premises, as well as a tightening of spatial controls: albeit only to new or converted premises. This apparent disjuncture between de-regulation and tighter controls, both to be exercised by the same local authority, is brought into tighter focus when the government's shifting stance towards local controls over licensing is examined in greater detail.

Local accountability and licensing control

The White Paper had proposed that local residents were automatically consulted about licensing applications and allowed to raise objections, alongside the police and other responsible authorities. A group of leading figures in the licensed trade were alarmed by this prospect and other aspects of the wholesale re-structuring of licensing. This group included leading representatives of pubcos and breweries. They argued that local authorities would be too responsive to the needs of the local electorate and would 'interfere' (McNeill 2001).

One of the leading operators went on to categorically deny that their pub chain was ever the cause of any extra disorder or violence. Giving evidence to a Parliamentary committee, the then executive chairman of J D Wetherspoon, opined:

> So many times we have had objections from the police or whoever to our applications and I have sat in courts and people have gone right through all the districts where we have opened up pubs to try to find evidence of an increase in disorder. There has been none. There is no evidence whatsoever that the opening of a Wetherspoon pub actually increases crime and disorder. It does not exist.
>
> (House of Commons 2003a: Ev 20)

The Licensing Bill moved licensing further away from local interests and gave more freedom to the licensed trade. Section 18 of the Bill stated that, in the absence of representations to the contrary, the licensing authority (that is the licensing committee of the local council) 'must' determine licences in favour of the applicant. This, and other proposed measures in the Bill, prompted greater organization on the part of residents' groups to form an effective lobby. The residents' groups, which were voluntary in nature, were supported by the Civic Trust, a charity dedicated to improving planning, building and the local environment and empowering communities. The Institute of Alcohol Studies is an independent charity that aims to act in the public interest by disseminating scientific knowledge about issues related to alcohol. An umbrella organization was formed between the two to coordinate activities and to influence legislation. Although the campaigners were anxious to highlight the harms caused by unrestricted expansion of licensed premises, they took a balanced view of consumption. Their evidence and lobbying recognized the values of expanding the evening economy in terms of vibrancy and jobs, but drew a distinction with the 'late night economy', that is, after midnight, which they regarded as threatening and intimidating and 'intolerable for long-suffering residents' (Open All Hours? Campaign 2002).

The provisions of the Bill were not changed and, consequently, local authorities were still charged with granting licences with whatever operating hours had been applied for, provided that no objections were lodged. The *Guidance* to the Act made clear that 'representations', that is objections to a particular application for a licence or to a change in licensing hours, should always be related to the four licensing objectives. It stated as a fundamental principle that 'local residents should be free to raise reasonable and relevant representations about the proposals' (Secretary of State for Culture 2002).

The explanation of 'reasonable and relevant' was that objections should only relate to the particular location for which the licence was being sought and that they were not 'frivolous or vexatious'. More amplification was given in the statement that any conditions that local authorities attached to a licence (for example that music could only be played during certain hours, or that customers could not drink outside) were to be 'necessary' and could not be 'inspirational'. This situation was undoubtedly an improvement for residents and local businesses that might have reason to object to a licence under the previous arrangements, because it ensured that their objections would be heard. Nevertheless, it still left objectors in a difficult position with regard to new licences because they would be making representations against a hypothetical situation with a difficult case to make about 'additionality'.

The Act also permitted residents another area of influence. This was in the power to call for a review of licence conditions, or of the licence itself if it could be demonstrated that the licensing conditions were not being met. Whereas a review of a licence previously had serious implications for a licensee because the only remedy was complete revocation and therefore potential loss of livelihood, the new Act allowed the licensing authority to impose lesser penalties. These included cutting back on activities or hours, or a temporary revocation of a licence while a new operating plan was drawn up. The focus on the individual licensed premises causing problems dovetailed with the interpretation of disorder being connected to the badly run venues, rather than recognizing the problems that unrestrained expansion had caused. As has been outlined, these 'negative externalities' are cumulative.

Given the constraints that had been placed on local controls in the Act itself, it was somewhat surprising that the rights of representation were given such prominence in a government press release announcing the passing of the Act. This explained that the Act would be 'responsive to the society it serves' and that it would offer 'a greater say for the public' (Department of Culture, Media and Sport 2003a).

While the press release was undoubtedly accurate, it nevertheless glossed over the limitations placed on residents with regard to their input into the process. Furthermore, it also omitted the salient point that previously a duty had been placed on licensing magistrates to act in the public interest and that a licensing magistrate could make their own objection to an application. By contrast, the local licensing authority was not permitted to object to an application on its own account. It was only a 'responsible agency', such as the planning department of a local authority, who could make objections on planning grounds, or another from a list that included the police and the fire service (Hunt and Manchester 2007).

Impacts of the Licensing Act 2003: hours

The Licensing Act 2003 became operational in November 2005. A transition period had been allowed to elapse before implementation to allow local authorities to consult on and then to adopt a licensing policy. Licensees had been allowed approximately six months to apply for licences under the new system. Before the Act came into force, some sections of the press reproduced and reinforced dire predictions of disorder: 'Our officers are overwhelmed by a sea of drunken, violent, vomiting yobs who, when they're not fighting each other, are falling through shop windows. That's now. What will it be like when we have a licensing free-for-all?' (Glen Smyth, chairman of the Metropolitan Police Federation in Slack 2005)

At the time of writing, it is only three years following this date and the full implications of the Act have yet to be realized. However, some observations can be made, based on empirical evidence gathered by the authors (see Chapter 1 for a discussion of methods used) and nationally gathered statistics.

Despite the political storm aroused by the extended hours permitted by the Licensing Act 2003, there is little robust evidence regarding the extent to which drinking hours have altered in England and Wales. According to the Licensing Minister (Department for Culture, Media and Sport 2007), only 3 per cent of premises hold 24-hour licences. Of this total, pubs, bars and nightclubs hold only 9 per cent, with the bulk belonging to hotel bars (65 per cent) and supermarkets and stores (18 per cent). The high figure for hotel bars might also be explained by some respondents including room mini-bars (Chair of Licensing, Westminster City Council, interview 2007). This relatively low number for 24-hour opening confirmed the intentions of major corporate operators as recorded in the authors' interviews prior to the implementation of the Act.

In March 2007, under one-third of licensed premises had authorization for 'late night refreshment activity', that is the provision of hot food and drink, either on or off the premises, between the hours of 11 p.m. and 5 a.m. (Antoniades *et al.* 2007). The DCMS report commented that premises do not necessarily stay open for as long as they have authorization. This was confirmed by the authors' own study:

> Yeah, we quite often hear in hearings where the application is 2 or 3 o'clock in the morning seven days a week. Quite often, we hear the

> manager say 'I have no intention of opening those hours but I'm being forced to make the application.'
>
> (Licensing officer and councillor, large town)

Establishing the extent to which premises are actually opening later than the previous terminal hours of 11 p.m. for pubs and bars and 2 a.m. for nightclubs is difficult. Case study evidence supported the ministerial view that the terminal hour for licensed premises in town and city centres have been extended for one or two hours at the weekend since the implementation of the Licensing Act 2003. This observation must be coupled with the fact that many venues had extended hours permissions before the Act anyway, so this represents a further, incremental change.

> Most of the pubs used to close at 11.00 and now close at 12.00 or 1.00, mostly at weekends. The clubs used to have public entertainment licences, which allowed them to stay open until 3.00, most of them stayed open until 2.00. The big clubs, the super-clubs and we've got two really big ones with 2,500... applied for and got very late licences but never used them.
>
> (Chair, Central Citizen's Forum interview, large town)

Impacts of the Licensing Act 2003: crime and disorder

A preliminary study of the impact of the Licensing Act on violence, disorder and criminal damage was reported by the Home Office in February 2007, based on the crime statistics in the year following the implementation of the Act (Babb 2007). Thirty police forces across England and Wales contributed statistics, with 18 providing detailed statistics for specific locations in city centres and near licensed premises. This group included Greater Manchester and London, but only the City of London and not the West End. The statistics for 2005–2006 were compared with 2004–2005. The overall conclusions were quite complex. Offences were grouped into five categories, with the smallest group, serious violent crime, making up 1 per cent of the total and the largest, criminal damage, comprising 53 per cent. In the two years surveyed, serious violent crime had fallen by 10 per cent, both during the day and the night, but with a slight increase between the hours of 3 a.m. and 6 a.m. Less-serious wounding followed a similar pattern, but with a 'particularly pronounced' rise between 3 a.m. and 6 a.m. There were small rises in the incidence of harassment and in assault with no injury, with a decline in harassment before midnight but a rise after it. Overall, criminal damage varied a little over the two years, but with a small rise during night-time since the Act's implementation.

Figures for the other key indicator of crime and disorder, alcohol-related admissions to accident and emergency departments of hospitals, are not available nationally (Elvins and Hadfield 2003). There are two contradictory detailed studies, with one investigation in the Wirral reporting a reduction in admissions for assault following the introduction of the Act (Institute of Alcohol Studies July 2007) and another study at St Thomas' in London, showing an increase in over-night alcohol-related admissions in the same period (Newton *et al.* 2007).

Further complications in interpreting the data are associated with different policing strategies and powers introduced before and after November 2005. Some police divisions had already introduced different policing tactics to cope with late-night disorder before the Act was implemented. Following full implementation, the Home Office also ran two Alcohol Misuse Enforcement Campaigns (AMECs). In addition, the police were given powers of arrest for assault with no injury and new powers to impose penalty notices for disorder (Babb 2007).

Despite the results being interim and the conclusions complex, there are, however, some observations that can be made. The first is that, contrary to the government's claims before the Act was introduced, the crime and disorder associated with late-night drinking has not been eliminated or nearly eliminated. Certainly, the reduction in serious and less-serious violent crime is welcome, but levels of disorder have not significantly changed. There also seems to be some evidence that there has been displacement later into the night. Case study material worked on by the authors provides some illumination to these observations. Prior to the implementation of the Licensing Act, an inner London borough undertook a detailed investigation of the night-time economy in various hot spots within its jurisdiction. The authors were able to analyse the data relating to three observation points in central London, adjacent to the West End.

Figure 6.2 maps the incidence of crime, fighting and other types of disorder and nuisance against the terminal hours of nearby venues. It confirms that fights were associated with the terminal hours of pubs and clubs. Other types of disorder continued into the later hours of the night. The impacts of staggered and later hours could therefore be inferred to reduce pedestrian densities at the peaks, and therefore reduce the flashpoints for fights and violent assaults but, nevertheless, push other types of disorder and low-level crime later.

Another example sets out this complexity. Before the Licensing Act 2003 passed through Parliament, Norfolk Constabulary and Norwich City Council were already taking action with regard to the problems associated with the dispersal of customers from licensed premises. The high levels of accidents on Prince of Wales Road (25 casualties per year, with over half occurring

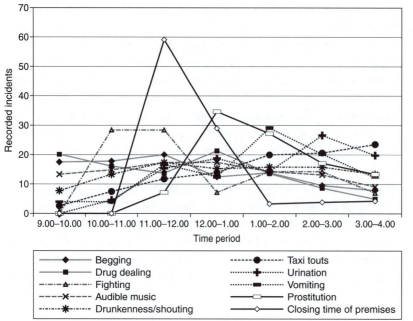

6.2
Terminal hour of 11 p.m. mapped against recorded incidents of crime and anti-social behaviour.
Source: Data derived from a study commissioned by a central/inner London borough, Crown Copyright, 2003.

at night) prompted the City Council and the highway authority, Norfolk County Council, to work with the Department of Transport in improving road safety. In addition to the road traffic problems, the council worked in partnership with the police to resolve some of the public order issues. Norfolk Constabulary launched 'Operation Enterprise' to take pre-emptive action and to 'impose an expectation of good behaviour'. Operation Enterprise was extended into the adjoining Riverside area for its second year in 2004–2005.

Although Operation Enterprise had ended by the time the Licensing Act 2003 was implemented, a dedicated team of police officers continued with its work and operational measures. They noted subtle, rather than dramatic, changes to patterns of dispersal. The main changes were a reduction in the 'pinch points' at 11 p.m. and 2 a.m. This has reduced pressure, both on the street and in queues for taxis.

> What is striking is that there's a very different rhythm to the evening. You could stand at the top of Prince of Wales Road and by 10 in the evening the road would be a sea of heads trying to get into clubs having moved from the pubs. That's not the case now, there are queues but they're not often more than 30 or 40 people.... People move on when it's reasonable or when they can get cheaper drinks. Again, at 2 o'clock in the morning, if you look down at the taxi queues there are two private hire offices on Prince of Wales Road ... I seldom see more than 20 or 30 people waiting at a time,

whereas previously you would see 100 people waiting and queuing there for 90 minutes after clubs closed.

(Police licensing officer, Norwich)

Despite the reduction in the numbers of people vying for services, there is still movement between the clubs. Once it becomes commercially unviable for a major club to stay open and they close their doors, the smaller clubs take 'other peoples' cast offs' for an extra hour or so until 3.30 a.m. This relieves the pressure of queues for transport and late-night fast-food. Nevertheless, this is not to say that all problems with dispersal have been solved. Lower-level public order offences in the late-night activity zones in Norwich have not reduced since the implementation of the Act, although serious violent crime has (Norfolk Constabulary, interview 2007).

A further important point, overlooked by many journalists but remarked upon by researchers in criminology and referred to earlier in this chapter, is that much of the less-serious crime and disorder are related to pedestrian density, that is to volumes of people and licensed premises. It is not, as the government previously argued, simply attributable to badly managed premises or a certain type of binge drinker. Babb (2007) suggests that the small rise in offences in the early morning is attributable to more people being out in the street through the extension in closing hours. The increases in pedestrian density are, of course, also related to the numbers, capacity and configuration of licensed premises. The relationship between these spatial and temporal variables, such as numbers, types of premises, capacity, terminal hours and their layout and associated violent disorder, is complex. In an international review of recent peer-reviewed research, Graham and Homel (2008) conclude that no direct causal relationships can be drawn. Nevertheless, in agreement with Hobbs' research team (2005a), they point out that increases in the density of licensed premises generally contributes to increases in violence and public order offences. The proviso is that isolated premises can also be troublesome.

These observations also help to explain why the Licensing Act 2003 has not had an immediate, dramatic effect on crime and disorder. The Licensing Act is essentially permissive and encourages a proliferation of licensed premises. The main arena for control lies in the conditions that are placed on operating hours and activities and in areas of 'saturation'. These are determined locally.

Impacts of the Licensing Act 2003: local accountability

The authors' study found evidence of improvements in the relationship between residents' groups and local licensees. There were some grumbles on the part of national operators in the on-license trade about the Act, but on the whole they felt that it was forcing them to take a more responsible attitude to neighbourly relations. Whether it was minimizing obtrusive music, cutting irresponsible promotions or forging better relations with the police or council, the majority of those interviewed felt the Act had generated a more accountable industry. These statements need to be tempered with the observations made by an independent team of researchers of irresponsible practices in a number of different locations (KPMG llp 2008).

The key difference that the Act appears to have made lies in defusing the adversarial nature of conflicts between licensees and local residents. Under the new regulations the system moved from a judicial process held in a court to a hearing in the local council offices. Although our study relied on only four case studies with five local authorities, chairs of committees were very aware of requirements to make fair and balanced judgements, and within the framework of national legislation and their own licensing policies. Despite the comments made by operators and, on occasion, the local police, all were emphatic that party political considerations and/or electoral advantage played no part in their decisions. In three of the locations, the committee chairs spontaneously commented on how the process of determining conditions had moved towards a negotiation between parties, rather than a contest. In the location that was smallest in size, the situation was slightly different and residents felt they had not been provided with sufficient information about the licensing process and nor had they formed into a residents' association. As a consequence, they felt disempowered. In contrast, in the most pressurized location, on central London's fringe, where there are a number of applications for new licences or for variations to operating hours, the process actually brought divided communities together. The Licensing Chair commented:

> The East End is quite a divided place. There's lots of communities living next to each other but are not interacting really except when it comes to objecting to a licence. People who would never really associate with each other from totally different walks of life will sit there and cheer each other on. It's nice [laughs] this unity of negativity. It's a striking thing. Also you get real vignettes of local history in these circumstances, so sub-committees can be really, really enjoyable affairs.

> (Senior councillor, London Borough)

Prior to the Licensing Act, licences were reviewed once every three years. Premises licences are now indefinite, but 'interested parties' and 'responsible agencies' have been given the right to apply for a review if evidence can be shown that any of the four licensing objectives are not being met, or if licensing conditions are being breached. Nearly 700 reviews have been held nationally, leading to the revocation or suspension of nearly 200 licences. The main outcome for the majority of reviews has been a change to conditions. Over 100 other establishments had their licensing hours modified (Antoniades et al. 2007).

A further point to be gleaned from this is that licensing policies vary considerably between different authorities. Even within the small sample of five authorities within our own study, it was striking that the smallest authority essentially repeated national guidelines, whereas the most sophisticated policy included a 'special policy' or cumulative impact area, together with more imaginative policies to encourage live music and performance. In fact, as of March 2007, there were 79 cumulative impact areas declared in England and Wales, of which 31 per cent were in metropolitan districts.

The national *Guidance* to the Act was meant to avoid significant local variations. Hunt and Manchester (2007), for example, have argued that the Licensing Act is politically a 'third way' piece of legislation (Giddens 1998), representing neither a 'top down' or 'bottom up' approach to regulation, but a hybrid between the two, with central government giving a strong steer to local controllers. In doing this, they point out that areas of confusion or variation have been created in the legislation, differences about which the national operators complained.

Impacts of the Licensing Act 2003: local diversity

The Licensing Act 2003 had the intention of enabling new types and varieties of licensed premises to be opened. The authors' study could find no evidence of this happening in the locations studied. Local council officials were clear that there had been no such diversification. Feminist groups, however, were noting an increase in diversity of a different kind. One unexpected and unforeseen consequence arising from the Licensing Act 2003 has been an increase in the number of lap-dancing establishments, doubling to approximately 300 (Object 2008: 5). The inclusion of 'erotic dancing' as a licensable activity meant that licensees could include this as one of a number of activities. Lap dancing includes different types of erotic dancing, such as table dancing, pole dancing, stripping and striptease, for a male audience. Many lap-dancing establishments describe themselves as 'gentlemen's clubs'. There are chains of 'adult entertainment' clubs operating at local, national and international level. The 'Secrets' chain has five clubs in London and offers 'fully nude table dancing and continuous stage entertainment'. Spearmint Rhino is an international brand with its headquarters in Las Vegas and branches in Australia, Russia and the UK. The 'mainstreaming' of erotic dancing is such that other establishments might not be dedicated clubs, but include pubs that have pole dancing or, say, striptease on Sundays.

Mainstream feminist objections to lap dancing are clear-cut, and in the UK a campaign has been mounted to seek a change in legislation that would have such 'adult entertainment' venues re-classified as 'sex encounter establishments'. The central rationale of the campaign is that lap dancing, striptease and other erotic dancing encourage and promulgate the objectification of women's bodies and runs counter to gender equality (see www.object.org.uk). Re-classification as sex encounter establishments would give local authorities greater control over the numbers and terms of operation of such venues. Although currently the licensing authority can impose conditions such as specifying that the dancer must keep a distance of 300 mm from members of the audience, in practice these stipulations are difficult to enforce. An undercover TV reporter found widespread disregard of such conditions in an investigation of lap-dancing clubs (Dispatches 2008). It is also difficult for a local authority to refuse a licence for these clubs because, as has been discussed earlier in this chapter, the only grounds for objection are contravention of the four licensing objectives. Generally speaking, lap-dancing clubs do not have the same problems with noise, unruly behaviour and so on as, say, a large vertical-drinking bar.

Previous small-scale research has provided evidence that there is a connection between the sale of 'soft' pornography and an increase in lap dancing, with an increased demand for the purchase of sex. One localized study found a 50 per cent increase in the number of reported rapes and charges of harassment in an area where a number of new clubs had opened (Eden 2007). Furthermore, many of the 'dancers' have to pay to perform and then make their money through direct payments. Under such stressed conditions there are strong temptations 'to go further' and have sexual contact with customers. There is also an association with people trafficking (Object 2008). Externally, clubs are not permitted to advertise with explicit images, but the blacked-out windows tell their own story and present a 'face' to the street that can be criticized on the same grounds as heavy metal shutters. That is, it looks hostile and, for the majority of women, unwelcoming. Clearly, the politicians who welcomed licensing reform as a way to 'shape the future of our towns and cities' did not have this scenario in mind.

Scotland

The Licensing Act 2003 applies only to England and Wales. Scotland has a different regime that has important divergences from the English and Welsh system. As in England and Wales, the Licensing (Scotland) Act (2005) was initially discussed in terms of tourism. The Act was introduced in response to the Nicholson Committee (Nicholson Committee 2003), which was established in 2001 to overhaul the previous 1976 Act. Scotland had seen a number of measures come into effect since the 1976 Act, and, as with England and Wales' Licensing Act 2003, the Committee's review and the subsequent Act was justified partly in response to what was perceived to be an unwieldy and overly bureaucratic system.

The new Act took effect on 1 February 2008. Like the English and Welsh model, the Scottish Act will see one premises licence and a newly introduced personal licence. The Act has also led to the creation of a new category of council officer, called a Licensing Standards Officer, whose powers rest between that of an environmental health officer and the police. Licensing Standards Officers are responsible for providing information pertaining to the Act, managing complaints and acting as a mediator in the event of any problems between parties. In order to acquire a licence, licensees must submit a plan to the Licensing Standards Officers and demonstrate how it accords with the five principles of the Scottish Act. These are: preventing crime and disorder; securing public safety; preventing public nuisance; protecting children from harm; and protecting and improving public health. This final principle is absent from the England and Wales Act. Exactly how this will manifest is, at the time of writing, unclear. However, the Act has resulted in several initiatives that may encourage and support more responsible licensing.

Under 'exceptional circumstances' licensees may apply for 24-hour licences. However, according to representatives spoken to in Scotland, terminal hours will not be significantly changed, but instead decided on a premises by premises basis. Under the current regime, casinos are able to open until 6 a.m., pubs and bars until 1 a.m. and nightclubs until 3 a.m. Objection, under the new Act, has also been updated with any person, irrespective of where they live in Scotland, able to object to a licensing application. In order to secure a premises licence, prospective licensees must first apply to the council in order to determine if the venue is suitable for a liquor licence. Local residents within a 4 metre radius will then be informed of a premises licence application within 21 days of receipt. Tighter controls on promotions have also been introduced, such as restrictions on special offers, which now have to be offered for at least 72 hours. This is designed to discourage happy hours or £10 'all you can drink' promotions, unless the licensee is willing to run such a promotion for three continuous days.

A significant component of the Scottish Act was the introduction of local Licensing Forums. Each local council must establish a Forum, and in large areas the council may establish more than one. The Forums are designed to monitor and review licensing in the local area and advise the Licensing Board. Importantly, the Forums are independent of the Board and are not allowed to comment on specific applications. Instead, they are designed to contribute to the overall discussion of licensing in the area. For example, members of the Forum may compile evidence of anti-social disorder in an area and present this to the Board as proof the Board needs to re-examine its licensing decisions. The Forum may also ask for a judicial review, such that their powers are both retroactive as well as proactive.

The Forums comprise only 5–20 people, representing licensees, the chief constable, the Licensing Standards Officer, a representative from the health, education or social work professions, a local resident and, significantly, someone to represent young people. In Scotland, this means the 16–25-year-old age bracket, and a young person over 18 can sit on the Licensing Forum. The distinction between the Licensing Board, which has power over licensing applications, and the Licensing Forum, who advise and monitor the Board, is especially important for residents. They are no longer required to comment on a specific licence in front of the licensee in a large and potentially intimidating environment. Instead, rather than attend the Licensing Board, a resident can instead speak to the Licensing Forum if they are unhappy with licensing decisions affecting their local environment. The Scottish Act, by virtue of the Licensing Forums, enables all residents to play a much greater role in the course of their local area.

A final component of the Act worth commending is the fifth provision, improving public health. This was absent from the English and Welsh Act, and though, at present, it is difficult to determine exactly how this will be played out (MP Consultancy 2008), its inclusion signifies a difference between the two Acts. Whereas the England and Wales Act was framed more in terms of tourism, the Scottish Act places health and tackling alcohol misuse at the very heart of licensing decisions.

Concluding comments

Licensing reform in England and Wales has turned out to be a missed opportunity. The emphasis on de-regulating permitted hours within the legislation was based on a narrow interpretation of the available research. Regrettably, the media backlash against the Act did not promote a more considered discussion of the problems of regulating late-night drinking, a discussion that should have started with the state of town and city centres that already had late-night opening. In retrospect, the spectre of 24-hour drinking has proved to be somewhat misleading, and extended hours – rather than all-night drinking – has been the norm. Licensing reform has had a positive effect on governance, however. It would appear that local authorities and interested residents can have an influence on the operation of the night-time economy in their area, although the circumstances in which they can restrict its overall growth are limited. In research conducted by the authors, it was found that licensees are having to accommodate and negotiate with local agencies and with local residents where they are organized. This process has become more transparent and accessible. Where residents and businesses are less well organized, they are vulnerable to changes that may not be beneficial.

Although it may be premature to be making a judgement, the main comments that can be made about changes in licensing legislation refer not to the changes that have been wrought, but to continuities. The impact on youthful binge drinking and low-level disorder appears to have been very limited. Even licensing hours have not undergone a dramatic change. The essential trajectory of UK drinking culture, a pattern that experienced a significant and rapid transformation in the late 1990s, continues. The supposed transition to a 'more relaxed style' of consumption has not been evident. It is to this continuity that we shall turn to next, seeking evidence from mainland Europe and further afield.

Chapter 7

Regulating licensing

The dream of a continental style of drinking

Bologna in Birmingham, Madrid in Manchester, why not?

BBC News 2003

The suggestion that changes to licensing regulation would lead to changes in British drinking culture has come to haunt the UK government. This chapter examines the promises made for a 'continental style of drinking' and the difficulties underlying a definition of a 'café culture'. It questions whether the idea of a perfect continent were mistaken from the beginning, based on a mythology about de-regulation that paid no attention to the laws and customs pertaining to other countries. These laws, as in Britain, encompass licensing, but also planning, building and environmental protection. In fact, many European countries have systems of control that are more integrated than in the UK and hence are more restrictive when seen as a whole.

In this context, an examination of arrangements for the disposition and regulation of licensed premises in mainland Europe is illuminating. There, many of the themes that have structured debate over licensing reform in the UK are equally pressing. These include expansion in the number of venues, increases in crime and disorder, clashes with residential uses, policy measures to promote tourism and the hospitality industries and concerns about under-age drinking and drug abuse. This chapter will draw on three case studies in order to examine this 'continental model'. Copenhagen and Barcelona are examples where public culture and drinking practices are widely regarded as a success story. Both cities are highly regarded in their use of public space. On another continent, both the Australian cities of Sydney and Melbourne have been faced with similar issues to the UK. Concern about binge drinking, anti-social disorder and under-age consumption have prompted intervention, marked by a rhetoric of 'continental' versus 'Australian' drinking culture and consumer preference. Before examining these examples, the intricacies of European practice and the 'European model', we will consider the way in which a 'European style of drinking' has come to the fore in media discussion of the evolution of British nightlife.

The Licensing Act 2003 and a 'continental style of drinking'

As has been noted, the first British government document to propose changes to English licensing laws had modest aspirations. In the passage from proposal to Bill, the goals of the legislation appeared to have grown from a reduction in bureaucracy to a more thorough shift in drinking culture. This change was associated with the proposal to completely de-regulate licensing hours rather than allowing an extension of one or two hours of extra drinking time. The White Paper explained the benefits of de-regulation in terms of tourists who, it argued, were often 'bewildered and confused' in Britain because they could enjoy a drink at any 'reasonable' time in their home countries. Restricted hours also caused inconvenience for others, tourists and inhabitants alike 'who want to eat or drink after seeing a film, attending the theatre or an evening concert' (Home Office 2001).

The explicit connection with Europe was made when a junior Minister stated that liberalization would 'eventually encourage the more sensible drinking culture seen in some other European cities' (Home Office 2001). The Regulatory Impact Assessment accompanying the Bill cited a prime benefit of the legislation to be boosting the tourism industry and 'improving our competitiveness with other European cities' (Department of Culture, Media and Sport 2003b). Licensing reform was going to achieve this through allowing better planning of towns and cities, encouraging greater diversity in choice of venue and in the age of the customer provided for and in allowing operators greater flexibility to provide what the public wanted 'when they wanted it'. Arguments about creating an environment where families could go out as a group and enjoy a relaxed drink are diffused throughout the document. Admirable as this may be, no evidence was brought forward to back up the claim that tourists were 'confused' or that de-regulation would necessarily lead to a greater diversity of offerings in the night-time economy.

Early in 2005, government ministers were pilloried for making associations between changing cultures and de-regulating licensing hours. The Prime Minister, Tony Blair, commented that theatre-goers would be 'inconvenienced' by being unable to secure a drink after the performance (Blair 2005).

This was later condemned by the *Daily Mail* as 'fatuous and feeble' (Anonymous 2005). The same comment had been headlined in the same paper three days earlier, juxtaposed with the photograph of a young woman apparently lying in a drunken coma on the pavement after a night out. The report's by-line read 'What planet do you think you are living on Mr Blair?'

Following this onslaught, ministers tended to be more circumspect in their claims.

A ministerial press release announcing the passing of the Act stressed the 'freedom and choice' that the Act would allow customers and claimed that a 'more civilized culture' in pubs, bars and restaurants would be produced by the ending of fixed closing times (Department for Culture, Media and Sport 2003). This argument was followed through in other interviews, although one minister broke ranks in November 2004, saying: 'We are trying to modernise the system and the industry is with us. We want to be more European. Like Italy or France' (Levy and Scott-Clark 2004). In turn, journalists conjured up an image of a café-culture allegedly present in mainland Europe to ridicule licensing reform. Sam Alexandroni, writing for a left-wing magazine and commenting on the continuing disorder following the Licensing Act 2003 noted that 'it's hard to imagine a scene more at odds with the continental café culture ministers hoped would flourish once drinking hours were extended' (Alexandroni 2006).

The constitution of a 'European' café culture is, of course, somewhat hazy. The practice of walking outside, in clothing that is intended to be viewed by others, to see and be seen and to meet and talk has been a long-standing custom in many southern European countries, and in Italy it is known as the *passeggiata*. Somehow, this has been conflated with the practice of 'circuiting', that is moving from bar to bar as on the English 'pub crawl'. Hence, in Degen's (2008) comparison of regeneration in Manchester and Barcelona, one of her respondents commented that the scene outside the bars and clubs in the Castlefields regeneration area was 'like Ibiza', with scantily clad people 'parading' around. The notion that licensing hours are unlimited on mainland Europe also permeates popular discourse; with an assumption that practices encountered within Mediterranean holiday resorts are somehow common to the whole of the European continent. In particular, the practice of eating and drinking outside, which is common to the tourist resorts of southern Europe, is often assumed to stand for the whole. Degen comments that 'the continental lifestyle theme fits into the picture of the 24-hour city simplified into the 24-hour consumption city' (Degen 2003: 876), a simplification that she terms 'economic continentalisation', and seems particularly apt.

To add to the complications of a more rigorous discussion of European practice, the impact of tourism itself cannot be discounted. Northern European visitors go back to their home countries from southern Europe and create a demand for a different type of public culture. An influx of foreign visitors, who 'read' the spaces of their chosen resorts and city centres as play spaces where normal constraints and standards of comportment may be discarded, play a part in the redefinition of local drinking practices. European Union policies and attitudes towards alcohol pricing, as discussed in Chapter 5, also produce convergence rather than differentiation. A 'snapshot' of the changing cultures in our two chosen cities follows.

Copenhagen: a southern European city?

Copenhagen is a city that set out 45 years ago to reclaim its streets and public spaces from the incursion of traffic. Prior to its adoption, there was no tradition of the intensive use of public space. Researchers at the School of Architecture in the Danish Royal Academy have extensively documented the reclaiming of Copenhagen city centre for pedestrians. In three intensive studies of the public realm in 1986, 1996 and 2005, they have charted the extent of the changes that have been wrought in the streets and public spaces within the inner city. This transformation has resulted in the type of external environment 'as we know it from more southern climates' (Kœbenhavns Kommune 2001). Gehl describes the horror with which the proposals for the first pedestrian scheme were greeted: 'We are Danes, not Italians' was the response. Nevertheless, the City Council pursued its policies.

These have achieved significant success. Traffic levels have remained constant at the levels reached in the 1970s, and the area given over to pedestrians has increased sevenfold since 1962 (Gehl and Gemzøe 2001). The city centre already had a relatively high number of residents, but this number has increased by 12 per cent over ten years with a housing rehabilitation and renewal programme. By 2005, 7,600 residents were living in the inner area of the city (Gehl et al. 2006: 28).

Copenhagen is also remarkable in the way that the City Council has actively encouraged a café culture. The programme of public-realm improvements incorporated encouragement for 'outdoor service'. This is strictly controlled, and café and restaurant owners have to pay for a licence to put out tables and chairs on selected months in the year. There are strict controls over the design and management of outdoor tables and chairs. The number of outdoor spaces has more than doubled over the last 20 years, despite the harsh conditions in autumn and early spring. In the inner-city area studied by researchers, the number of outdoor café chairs increased from 2,970 in 1986 to 7,020 in 2005 (Gehl et al. 2006: 41). The increase has meant that many areas in the city centre have now become 'café quarters'. Although the phenomenon of the 'continentalization' of Copenhagen is based primarily on cappuccino, night-time activities have played a significant part in the growth of public life in the city. In the summer one-third of time spent outdoors in public spaces is in the evening (Gehl *et al.* 2006).

The growth of Copenhagen's café culture has been facilitated by the Danish planning system. This permits shops (retail) to be turned into cafés or restaurants, without requiring special planning permission. There is an upper limit on the total floor area of retail space within particular areas within the city and the policy has been to limit the amount of retail with a view that demand will be diverted to the nearby new town of Orestad. These limits prevent a pro-liferation of cafés and night spots. In addition, the planning system can control the location of large late-night venues. As a City Council planning officer remarked, 'there are no restrictions on changing shops to bars, restaurants, etc. but you cannot change retail into banks, clinics, etc. The aim is to maintain a living environment, not "dead" uses' (interview 2001).

The cultural division of the City Council controls liquor licensing. Applications are made to the licensing division and are determined by a Licens-ing Board. The Licensing Board has 12 members who comprise local politi-cians, police officers and local authority officers. Interestingly, there is no notice procedure within Copenhagen to publicize licence applications to local residents or to members of the public. There is no right to object prior to the granting of a liquor licence. Applications are considered and determined in private. There is no hearing, no right of representation and no right to appeal. As one council officer remarked to us in 2001: 'If there were a right to appeal, then everybody would do it.' Instead, the police carry out an investigation and particular atten-tion is paid to the status of the applicant and to the business plan put forward for the premises. This has to be viable and the regulations forbid financial backing from a beverage producer. Environmental protection officers also make a report with recommendations. Their report considers local impacts such as noise, smells from cooking food, proximity to neighbouring flats and houses, closeness to other licensed premises and the availability of transport.

The Licensing Board, on the recommendation of the officers, deter-mines terminal hours for night-time activities for each licensed venue. Gener-ally, if the venue is in a residential area, then the hour is set at 1 a.m. If it is in a mixed-use area, with flats, offices and retail and entertainment uses, then the licence can be until 3 a.m. or 5 a.m. As an environmental protection officer remarked to us in 2005: 'If it's 5.00 a.m. then in Denmark it's really 24-hour drinking.' Residents' complaints about noisy music from venues are processed by the environmental protection officers and the police who can both impose fines. No music is allowed in the street, apart from during special events, such as the jazz festival. The controls over noise levels emanating from venues are also strict, with a regulation that limits intrusion into neighbouring dwellings to 30 dB before 10 p.m. at night and only 25 dB after that. Noise limiters on sound equipment within venues are set at 80 dB, a slightly lower limit than the UK. Problems with disorder outside venues are the remit of the police. If there are complaints about infringements of licensing conditions then the police have the ability to investigate, using powers conferred by local by-laws. They are able to act quickly, imposing fines and, in extreme cases, closing down venues.

The police and the City Council have used the measure of extending terminal hours to dispel the problem of 'circuiting', that is, customers moving from one venue to another later in the night. These extensions have been applied in a controlled manner, with care taken not to allow too many venues to operate in clusters.

> There were discussions about staggered opening hours – the police used to have problems when most venues closed about 2.00 a.m. and then people would walk around looking for a place that would be open till later. So now we give most places till 5.00. There is a new thing now that people stay in their houses until midnight and only then go out. We allow many 5.00 a.m. licenses but we always make sure that the rules are fulfilled. We try to make sure that places are not too close to each other, because you get an area just like Nyhavn, whereas in other places if it's only bars, you get areas just like in Hamburg. We don't want that.
>
> (Environmental protection officer, Copenhagen City Council)

The area referred to by the environmental protection officer is a well-known micro-district in Copenhagen (Roberts *et al.* 2006). Nyhavn is an area on the fringe of the city centre, surrounding a canal dug in the seventeenth century. Attractive merchant houses line the sides of the canal and its wide quaysides. By the 1970s the area, which had emerged as a red-light district in the preceding century, was facing gentrification. The local residents' association took action, with the support of the City Council, to conserve the buildings and maintain the small resident population and the remaining shipping that used the canal. Nyhavn was reinvigorated as a tourist attraction, the quaysides were pedestrianized, licensing relaxed and outdoor tables and seating encouraged.

The area quickly became gentrified and established itself as one of Copenhagen's main tourist attractions. The newly restored houses, most of which are now listed, were converted to warehouses and flats, causing rental prices to increase and some of the original residents to move out. The 'sunny side' of Nyhavn became an entertainment strip, saturated with bars and restaurants; as the manager of the Nyhavn Business Association (*Erhvervsforening*) admitted, 'we cannot have more restaurants. Every building on the sunny side has a bar or a restaurant.' Economic pressures asserted themselves and each of the premises started to use their space more intensively, selling more alcohol to boost returns. The neighbourhood started to suffer from problems with drunkenness, excessive noise and litter.

To counteract these negative effects, the Business Association brought financial clout to some of the residents' initiatives, such as a new bridge to span the canal and a new waste disposal system. This uses a central suction system to take the rubbish out into lockable waste ducts 500 m away, where it is more easily collected and disposed of. The business and property owners, together with the residents' association, raised a substantial sum to do this and were supplemented by the municipal authority. In addition, a new public toilet was installed and initiatives such as holding a Christmas market initiated to re-popularize the area.

More recently, Nyhavn has been the subject of a further experiment. Influenced by Richard Florida's research (Florida 2002), both the Greater Copenhagen Authority and the City Council have sought ways in which to expand the creative industries and cultural economy in Copenhagen. A consultant's report recommended de-regulation within the night-time economy, with fewer constraints on the use of public space for concerts and events. It was also argued that it should be easier to open nightclubs, bars and restaurants, and shops should be encouraged to stay open later. While the report recognized that this would lead to a 'more raucous' city, it was argued that the move towards a 24-hour city would attract sufficient creative professionals and industry for this to be worth-while (Bayliss 2007).

In 2006, Ritt Bjerregaard was elected the new Lord Mayor of Copenhagen and she has been willing, thus far, to put some of these ideas in place. A council official told us that Bjerregaard's idea is that Copenhagen should evolve in a more metropolitan manner, with a response to the criticism that the city is a 'little bit small town' in its attitudes towards activities in the early hours of the morning. The impetus is to boost tourism, from within Denmark and outside. The main competitors are seen to be Sweden and the north German cities of Berlin and Hamburg. Each of these is reputed to have an unrestrained nightlife with concentrations of bars and clubs.

The mayor has the twofold objective of wanting to make life easier for operators in the city, and at the same time the City Council wants to encourage more residential uses in the city centre. The Environmental Protection department was asked to resolve the two divergent objectives. As a start, the Department suggested a small experiment in Nyhavn. The City Council decided to allow outdoor service at the café tables until 2 a.m., a two-hour extension. This experiment was tried in the summer of 2007 and will be repeated in 2008. The evaluation is to be in September 2008. The residents of Nyhavn objected because they were worried about noise, violence and disturbances. However, this small experiment is a 'big event' in Copenhagen and much talked about in the local media. At the time of writing, the Environmental Protection Department 'have not learnt very much' because the summer in 2007 was exceptionally wet and most premises could not sustain an outdoor service (interview 2007).

Meanwhile, it has been made easier to open nightclubs, bars and restaurants in the city centre. However, the noise restrictions make building operations expensive and this inhibits expansion. A further hindrance to expansion is the increasing return of commercial properties to residential uses. Once a residential use is established, neighbouring licensed premises have to comply with the noise regulations and entrepreneurs are reluctant to take a risk on premises where this type of adjacent conversion might happen.

In conclusion, Copenhagen has transformed itself from a rather dour northern European city to one that enjoys a genuine, southern European style of outdoor café culture. Its transformation supports Stephen Green, the Chief Constable of Nottinghamshire's comment on the 24-hour city that: 'If we want a Continental café culture – build cafés' (Taylor and Hughes 2005). The restrictions placed on the development of the night-time economy in Copenhagen have assisted in maintaining a balance between 24-hour drinking and the rights of residents to enjoy a period of quiet. This balance has not been achieved smoothly and problems familiar to the UK in terms of 'circuiting', crime and disorder have been dealt with through a judicious allowance of later hours in a controlled number of venues. Apart from in Nyhavn, night-time venues are evenly spread through Copenhagen city centre.

The desire on the part of the City Council to loosen Copenhagen's regulatory structure in order to compete more successfully with its close neighbours, Berlin and Hamburg, reflects a global pressure. So far the experiment has been rather modest. When the environmental protection officers and the city planners revise their municipal plan and try to achieve the twin objectives of de-regulating nightlife and attracting residential uses into the city centre they might care to consider the experience of Barcelona. It is to this city that we shall turn to next.

The night-time economy and the Barcelona 'model'

Urban planners lauded Barcelona in the 1990s as the 'model' for urban regeneration. Barcelona's successful reconstruction was a result of a good deal more than an imaginative programme for the renewal of nearly 200 public spaces throughout the city, and the bold but morphologically sensitive works associated with its hosting of the 1992 Olympics (Roberts 1998). Marshall (2004) reminds us that planning in Barcelona after the death of the Spanish dictator, General Franco in 1975, has been multifaceted. Interventions, projects and strategic plans have skilfully combined physical changes with cultural and eco-nomic strategies, underpinned by temporary political alliances between differ-ent tiers of central, regional and local government and autonomous residents' associations. It was this convergence of physical interventions, economic success and political consensus that provided the rationale for the Royal Institu-tion of British Architects to award the city of Barcelona its annual gold medal in 1999. The city was also an important reference point for the Urban Task Force's *Towards An Urban Renaissance* (Urban Task Force 1999), which encouraged British cities to adopt a more continental style of urban form. Extensive aca-demic discussion and analysis published after the millennium, however, has pointed out the historical specificity of Barcelona's success and provided evid-ence that the city is currently experiencing problems with its planning and regeneration programmes similar to other cities in the northern hemisphere (Balibrea 2004; Garcia-Ramon and Albet 2000; Monclus 2003).

There is another aspect to this discussion that is both relevant to the night-time city and comparatively under-researched. In a different interpre-tation, based on urban design theory, the Barcelona 'model' is that of a compact, 'vibrant', mixed-use city with a network of attractive streets and public spaces. Its inhabitants, who are housed at what are, by European and North American standards, breath-taking densities of 400 dwellings per hectare (Urban Task Force 1999), want to live there and are proud of their city. Of course, this model of high-density living applies only to the central core of Bar-celona, specifically the Old Town (*Cuitat Vella*) and the 'extension' (*Eixample*), with its incorporated neighbourhoods such as Gràcia. The outlying suburbs and towns in the greater metropolitan area are at lower densities and are more functionally segregated. Within the central core, however, the adage of being able 'to work, rest and play' within the same physical neighbourhood is still per-formed by approximately half the inhabitants in the metropolitan region (Monclus 2003: 415).

For visitors to Barcelona and perhaps especially for emergent urban professionals on their undergraduate field trips, the central areas of Barcelona provide an example of a compact city form that appears to easily accommodate the fabled southern European lifestyle. Such visitors enjoy the lively public spaces and the 'vibrant' nightlife that provides a startling contrast to the sani-tized suburbs of northern cities. The Urban Task Force report argues that the density of the city, combined with the character of the public realm, with its network of public spaces, layout and connectivity 'plays a fundamental role in linking people and places together', thereby ensuring social cohesion (Urban Task Force 1999). As with other aspects of the Barcelona 'model', circum-stances are changing rapidly and, while this observation may have applied in the earlier years of Barcelona's regeneration, there are many tensions and difficulties. These tensions revolve around familiar issues in the governance of the night-time economy; for example, noise nuisance, licensing control and standards of acceptable behaviour in public spaces. In the paragraphs that follow, some measure of the tensions generated by economic expansion and high-density living will be explored, based on a small number of interviews with City Council officials, representatives from external agencies and inhabitants.

The city's regeneration and re-imaging strategies have boosted its tourism, with overnight stays increasing from 3.8 million in 1990 to 13.2 million in 2006. Barcelona is now the fifth-most visited out of ten major tourist cities in Europe. The increased number of visitors to a city whose centre already hosts a population of 1.6 million, within a wider metropolitan area with a further 1.5 million inhabitants (Barcelona Turisme 2007), has increased the demand for restaurants, clubs and bars. Despite the City Council's attempts to promote itself as a destination for art and architecture, sport and heritage (Smith 2005), the city attracts visitors who come for its nightlife alone. A worker in a drug prevention scheme explained:

> One of the problems is when English people come for their stag parties and a group of fifteen girls or boys coming over on a cheap plane and staying over and not even staying in a hotel for one night. They are coming for one crazy party making a lot of noise, staying one night in the street or whatever and going home. You have agencies selling these 'Ibiza' tours with even drinks included in the package.
>
> (Drug prevention officer, Safer Nightlife Project, Barcelona)

As with other European cities, nightlife provision is controlled through the regulatory regimes of planning, licensing and environmental protection, with enforcement provided by the police. With the influx of tourism, various neighbourhoods in the city centre boomed. The historic core around the cathedral, the Barri Gotic, had a neighbourhood plan, a PERI (*Plan Especials de Reforma Interior*) that protected it from too much expansion in terms of 'hospitality-related' land-uses (Esteban 2004; Garcia and Claver 2003). To this day, the neighbourhood provides a delightful selection of small specialist shops, together with small neighbourhood bars, cafés, coffee shops and restaurants. Other neighbourhoods and micro-districts were less fortunate. The medieval core of historic Barcelona, with its narrow, winding streets and alleyways, offered almost the perfect setting for nightlife, with its reputation for prostitution and drugs providing an 'edginess' and marginality associated with transgression and liminality. Local residents, however, were less impressed by an incursion of 'designer' bars, 'hip' shops and restaurants that replaced their local bars, greengrocers and everyday retail outlets. Residents in the Ribera on occasion threw eggs at noisy revellers, displayed home-made placards demanding the right to sleep and unfurled banners objecting to their streets being used as open-air toilets (Chatterton and Hollands 2003). Residents voiced criticism of the City Council's willingness to grant new licences and argued in a newspaper article that 'the leisure culture is now incompatible with the residential needs of the area' (Chatterton and Hollands 2003; Degen 2004: 140).

Degen has movingly described how one of the poorest areas in historic Barcelona, el Raval, has been re-created as a cultural quarter, focusing on the iconic Museum of Modern Art, known as the 'MACBA', designed by the renowned architect, Richard Meier. A new network of public spaces has been inscribed into el Raval, extending and developing the historic street pattern in the north, as in the Plaça dels Angels, and in the south, following a nineteenth-century model of slum clearance with a newly cleared Rambla. Degen demonstrates that rather than providing a renewal of the social cohesion discussed in the Urban Task Force Report, the gentrification attached to these public-realm improvements has resulted in a layering of space that has shattered the pre-existing rhythms and homogeneity of its culture. On occasion, this conflict between the influx of visitors to the cultural attractions and the 'cool bars' that surround the MACBA erupts into 'place wars' (Degen 2008). The influx of people from all over Barcelona at the weekends causes local residents to protest at the noise and disruption, while at different times of the day school children and young adults reclaim the newly designed spaces with their skateboards and mopeds. While it might be thought that binge drinking is purely a British phenomenon, a recent study demonstrates its increase in Spain, particularly amongst higher-income young men (Valanecia-Martin *et al.* 2007). As was also explored in Chapter 5, the more recent botellón phenomenon, where Spanish under-age drinkers meet in public areas to consume alcohol, attests to how the European ideal does not translate into an absence of alcohol misuse (Gual 2006).

It is in the old district of Gràcia that 'place wars' have become most acutely expressed around nightlife. Gràcia is a district of Barcelona that was autonomous before the nineteenth-century 'extension' of Cerda's famous masterplan. It is less dense than the Cuitat Vella, but nevertheless provides a series of streets and spaces laid out in a distorted grid pattern that provides charm and variety. In the early 1990s it accommodated Barcelona's normal ratio of bars, clubs and restaurants. In the decade leading up to 2003 this figure increased dramatically, to approximately 800, in an area with approximately 70,000 inhabitants. The manager of the remaining alternative bar cum community centre, run by a left-wing collective, told us that rents had increased too, from €600 (US$750) to €2,000 (US$2,500) per month for a typical bar in 2008. These increases have not only driven out some of the traditional crafts and artisan workshops from the area, but have also all but excluded the leftist collectives, squatters and bohemians who previously formed part of the community.

While the remodelling of the Plaça del Sol has been lauded in urban design textbooks (Gehl and Gemzøe 2001), the expansion of nightlife has resulted in a loss of vitality around the square during the day. None of the ten bars that surround it are open in the morning. Meanwhile, the noise of revellers partying until the early hours caused one resident to throw an iron bar out of the window in 2004. A local councillor took action through the formation of a Forum del Silenci, and a new residents' association was formed to combat noise and uncivil behaviour (BCN Living 2005). It was also in Gràcia that the owner of the El Portet bar-restaurant so enraged his neighbours that he found himself a recipient of a four-year jail sentence in March 2006. In fact, the owner had been operating the premises illegally, for it had been refused a licence and closed down five times over a two-year period because of insufficient sound insulation protecting residents in the neighbouring building from the sound of the freight lift, the mechanical ventilation and the extractor (*Barcelona Reporter* 2006). The court awarded damages to two neighbouring households for the costs of their medical treatment following the sleep deprivation they suffered.

Following the rising tide of complaints from residents across the city, the *Ajuntament* or City Council decided not to issue any more licences in 2003. Its planning department is drawing up a new 'Civil Action' plan for the ten neighbourhoods in central Barcelona. These are currently subject to consultation and the proposal is for implementation in 2009–2011. The intention of the plan is to provide a liveable mix of different activities throughout the city, with no concentrations of licensed premises. As a City Council official explained:

> one of the key points particularly with the council is to avoid the segregation of the city into areas of housing, areas of business and areas of night life, to have a mixed city. We know this policy does produce problems, as you have to try to manage the cohabitation between nightlife, leisure and people resting, and neighbours. In terms of quality of life we prefer to keep that within the city and have controls over it. Of course, it's a better option, rather than expelling it [nightlife] to industrial zones outside the city, that become isolated from the city and problems increase in terms of accidents, traffic problems, people drink driving and also criminals, criminality and so on. This is the council's municipal plan.
>
> (Representative, Department of Community Safety and Mobility, Barcelona City Council)

The plan will set out percentage limits for particular types of venue, with a recognition that the maximum limit has already been reached in some places. Planning rules also set out distances that particular types of uses have to be from each other. For example, a music bar cannot be next to a hospital or an old people's home. The intention is to avoid specialized zoning, within entertainment itself, to avoid, for example, a 'zone of bars'.

In this approach attention is paid at the level of detail of the street. A City Council official explained:

OFFICIAL: The Civic Youth Plan is to impede excessive concentration of certain types of premises in certain areas.

INTERVIEWER: What types? Is it big bars and discos?

OFFICIAL: They are just as worried about the big venue in one place as lots of little bars in one street. It could be a music bar, restaurant and discos. The disco closes at maybe 5.30 in the morning and the bar opens at 6 in the morning. People go straight to the bar and they are doing the same thing, the bars, opening half an hour before. You try to have a bit of everything but order it more. It's just as important what happens in one street as in the big venue.

Because operating schedules appear to be so long in Barcelona, it might appear that licensing regulations are much looser than in other European countries. This is the reverse of the truth. There are ten types of licensed premises (*locals*) specified in the regulations. Some of these relate only to licensing arrangements that are no longer used. Table 7.1 sets out the most relevant contemporary categories. Problems occur when normal bars transform themselves into music bars without seeking a change to the category of their licence. Licensed premises also have strict regulations about their '*aforo*', or capacities, which have to be displayed within the premises. Following a nightclub fire in Madrid, all Spanish nightlife venues now have strict regulations with regard to security and fire. Enforcement has been strictly pursued in recent years and has aroused controversy, not least from the association of nightlife operators, FECALON.

Table 7.1 **Selected categories of licensed premises in Barcelona and permitted hours**

Type	Earliest opening hour	Latest closing time	Extra time for Fridays, Saturdays and festivals
Discotheques	17.00	05.00	1 hour
Sales de fiesta	17.00	05.00	1 hour
Cafés cantant	17.00	04.30	30 minutes
Cafés concert	17.00	04.30	30 minutes
Cafés teatre	17.00	04.30	30 minutes
Bars musicals*	17.00	02.30	30 minutes
Bars C1 and C2	06.00	02.30	30 minutes
Bars-restaurants	06.00	02.30	30 minutes

Source: Derived from (FECALON, 2007).

Notes
*These bars have more expensive licences and more restricted hours than bars C1 and C2. There are proposals to allow live music in *Bars musicals* in order to increase the differentiation between the categories.

Problems also arise in Barcelona with under-age drinking. Spanish law permits 16-year-olds to enter licensed premises but not to drink. In crowded premises, this becomes difficult to monitor and its is the owner of the nightclub who is held responsible should the police discover under-age service. Enforcement is complicated due to the four different categories of police operating in Barcelona. In addition to the two types of national Spanish police, the national police and the Guardia Civil, there is a newly formed police force run by the Catalan Regional Council, the *Generalitat*. The municipality also has its own force. Both forces seem to operate to enforce licensing controls, which annoys licensees. From the point of view of the City Council, enforcement is also a minefield. In Barcelona, in contrast to the UK, large chains do not dominate the ownership of nightlife. No single owner has control of more than 1 per cent of licensed premises. It is, however, the arrangements for responsibility for the licence that causes the City Council problems. Frequently the title-holder for the licence is not the same as the business who is running the premises, thereby subverting enforcement measures:

> you had a situation the owner would rent [the premises] out to somebody and they would set up parties, noise, all kinds of problems, uncontrolled consumption. The person responsible for the action was the person renting out the property. So the fines would be placed and maybe that particular venue was closed down but then the licence was still there and then they could open a new venue. The person had the licence to rent it out again then you start over. The council has to start from zero as there is nothing against it, in reality you could do this indefinitely as it's rented out as different premises.
>
> (Representative, Department of Community Safety and Mobility, Barcelona City Council)

To combat this difficulty the City Council is now pursuing the licence holder as well as the business operator. The authority has been active, and in 2006 awarded closing orders on 256 licensed premises in the city. The majority of these were for infringements of licensing hours (28.3 per cent), for permitting unauthorized activities (23.3 per cent) or simply for operating without a licence (17.4 per cent) (FECALON 2007).

It is not only bar owners who are experiencing the force of the City Council's enforcement procedures. To combat anti-social behaviour in public places, a new series of municipal by-laws (*Ordenança*) have been adopted (Ajuntament de Barcelona 2006). These are a more tightly drawn set of rules governing behaviour in public places, formulated in response to the city's expanding visitor economy:

There was that problem; the question of skate boarders, rollerblad-
ers, low cost tourism and people didn't even go in a hotel, just slept
in the street. Stag parties and hen parties, although they went to
other cities, all those things were putting a massive pressure on
public space, that added to the local things, beggars, prostitutes and
international networks, then with the trade of women from Africa
and eastern Europe for prostitution which produced a problem in the
city, things weren't right.

(Representative, Department of Community Safety and Mobility,
Barcelona City Council)

The intention of the by-laws is to restore peace and 'dignity' to public space.
The by-laws control many types of activity, from prostitution and drinking and
selling alcohol through to fly-posting (bill sticking) and the chaining up of bicy-
cles. The measures were passed without a political consensus, with the social-
ists and Catalan nationalists in favour, but the ex-communists against. A City
Council official explained that it was understood that the by-laws would not
eradicate problems with contestation over public space, but that they would
'set limits and give a message' so that people could sit down in public space
without harassment. This official thought that the measures had achieved 'a
certain culture of civic behaviour and responsibility'.

Penalties for infringement of the by-laws are quite steep. For
example, one member of a leftist group complained of a €300 (US$375) fine for
sticking a poster on to a tree with adhesive tape. Residents within the metropolitan area are also annoyed that visitors appear to have been exempted from the
legislation. The spectacle of Scottish football fans urinating in the fountains of the
Plaça Catalunya, the principal square in the city, beneath a commemorative
statue of a Catalan hero and not being charged with any offence caused an
outcry in November 2007. According to a newspaper article civic leaders were
'deluged' with complaints and the police came in for 'bitter criticism' (Hunter
2007). Apparently, the police regarded the match as an exceptional circumstance
and decided not to enforce the by-laws. Councillors have promised enforcement
on visitors in the future, but face difficulties because under Spanish law, it is prohibited to confiscate passports to prevent the offender from leaving the country.

Further proposals to deal with the negative externalities derived
from Barcelona's night-time economy include a proposal to allow larger
premises, discotheques and *sales de fiesta* to stay open for an hour longer on
weekdays. This is to prevent disturbance when all-night revellers move from
the clubs at 5.30 a.m. to the early-opening bars, which only open half-an-hour
later, at 6 a.m. There are also 'platforms', that is cooperation between the
police, drug prevention agencies and the licensees to raise standards and
control drug and alcohol misuse.

This brief excursion into the details of Barcelona's nightlife, its negative externalities and its regulation, explode the myth of a 'relaxed drinking culture' that arises through a mature licensing regime that permits long hours of operation. The economic pressures on Barcelona's nightlife have produced similar problems with noise and other nuisances, anti-social behaviour, alcohol and drug misuse as compared to town and city centres in the UK. Problems associated with concentrating licensed venues and dwellings in close proximity are not resolved purely by some erroneous notion of cultural or national temperance. Although in Spain there are indeed different patterns of drinking practice, and different levels of consumption, it is not enough to presume these contrasting practices will resolve what are otherwise spatial and temporal planning issues. Nor can the corporate ownership of licensed premises be held to be the root cause of unlimited expansion or alcohol-related disorder (Hollands and Chatterton 2003). Barcelona's problems have occurred in a context dominated by small- and medium-sized enterprises. Some respondents argued the problems were particularly acute in Barcelona because of its soaring housing costs. Young people were unable to form independent households of their own, yet had reasonably high disposable incomes, which creates a 'push' factor towards 'going out'. This seems a reasonable explanation and also applies to many places in the UK.

This comparison between Copenhagen and Barcelona has highlighted the difficulties surrounding creating or sustaining a 'European' style of civilized and restrained consumption in an era of international tourism and transnational competition. It is highly pertinent that the Licensing Act 2003 was passed in the UK at the same time that the city of Barcelona decided to withhold further liquor licences. Within mainland Europe, different cities, regions and countries have adopted a variety of different regulatory regimes to control common issues regarding under-age drinking and responsible service, the proliferation of venues, size and style of premises, their capacity and impact (Boella et al. 2005; Roberts et al. 2006). Debates that draw on a notion of a 'continental style' of drinking are not confined to the 27 member states of the EU, however. In the section that follows, we shall consider the impact of such debates on another continent.

Australia

In Australia concern about binge drinking, anti-social disorder and under-age consumption have prompted an increased tax on alcopops, and for the Australian Prime Minister, Kevin Rudd, to launch a series of initiatives to tackle alcohol misuse (Rudd 2008). Against this backdrop, New South Wales (NSW), in which the state capital Sydney is located, overhauled its licensing regime in 2007, with the introduction of civilized 'small bars' and licensed cafés serving as a cornerstone of the legislation. Melbourne, the capital of the state of Victoria, served as the inspiration for NSW's legislative changes. Nonetheless, debate about the proposals were marked by a rhetoric of continental versus Australian drinking culture and consumer preference.

In Australia, each state is responsible for their own licensing regimes and has its own government structure and laws. In NSW, a new licensing act came into effect on 1 July 2008, the Liquor Act 2007. As was the case in England and Wales, the NSW legislation was discussed in terms of ena- bling a more continental style of drinking practice, one that would hopefully herald a more sedate and civilized form of consumption. It was not mainland Europe that served as the model, however, but Melbourne. Widely hailed as the most 'European' of Australian cities, Melbourne has acquired a reputation for offering a diverse and cosmopolitan bar and café culture. The city is often cited as the third-largest 'Greek' city in the world, with the number of Greek speakers third only to Athens and Thessaloniki. Its late-night culture is also diverse and marked by traditional pub-hotels, as well as late-night licensed cafés and small designer bars. These small bars, in particular, have drawn a great deal of praise for not only their contemporary and often innovative design, but also for enabling a relaxed form of sociality and drinking practice.

Melbourne's bar scene has been widely attributed to licensing laws introduced by the former conservative Premier for Victoria, Jeff Kennett. Kennett, when in office between 1992 and 1999, overhauled the state's licens- ing laws, resulting in the current Liquor Control Reform Act 1998. The aims of the Act included '[to] facilitate the development of a diversity of licensed facili- ties reflecting community expectations' (Liquor Control Reform Act 1998: 16). This objective, 'a diversity of licensed venues', is attributed to the growth of Melbourne's small-bar scene and from which NSW drew inspiration. Victoria's Act, first and foremost, overturned the prohibitive costs for licensing cafés and restaurants, thus allowing operators to serve wine or other alcoholic drinks to patrons. Entrepreneurs were able to purchase a licence more readily and eco- nomically, which in turn resulted in the development of small bars. Melbourne's success in revitalizing its late-night culture, however, can also be attributed to a number of other interrelated spatial, social and economic factors. The city's small-bar culture developed principally in its formerly under-utilized lanes and alleyways, of which there are approximately 180 dotted around the central busi- ness district. The lanes, which are also home to a number of small cafés, bou- tiques and galleries, have been integral to the city's broader urban renaissance (City of Melbourne and Gehl Architects 2005) and the manner in which Mel- bourne has fashioned itself as Australia's premier cultural and style capital.

By reducing the cost of a licence, Victoria's laws allowed for a new type of drinking venue to flourish, but they have not led to the complete eradi-

7.1
A typical 'lane' in Melbourne, Australia.
Source:
Photograph supplied by Ainsley Crabbe.

cation of alcohol-related violence and misuse. Melbourne's central business district is believed to attract up to 300,000 people on a typical Friday night (Eltham 2008) and the media in Victoria, just like the neighbouring state of NSW, has recently focused heavily on late-night disorder and binge drinking. As of 2008, a statement of policy was introduced which sets a 'freeze' on all new licensees applying to operate past 1 a.m. This affects the cities of Melbourne, Port Phillip, Yarra and Stonnington.

Despite the continuing problems with alcohol-related disorder and misuse, Sydney is now hoping to follow Melbourne's lead in encouraging small 'owner-operator' bars to open. The NSW Liquor Act 2007 (NSW Office of Liquor, Gaming and Racing 2007), supported by the Liquor Regulation 2008 (NSW Office of Liquor, Gaming and Racing 2008) is owed in part to Clover Moore, who acts as both an Independent member of the NSW Legislative Assembly and as Lord Mayor for the City of Sydney. In September 2007, Moore introduced a private member's Bill that sought to amend the existing Liquor Act (1982). Until 2007, liquor licensing in NSW was decidedly complex, with many categories of liquor licences for different types of premise. Under the 1982 Act, NSW had drink or dine laws which effectively kept the two forms of leisure separate. Restaurants wishing to serve alcohol only were required to hold a licence that could cost up to AUD$15,000 (US$9,500). The restaurant would also have to continue serving food to at least 70 per cent of its patrons. Fees for premises licences varied widely, with a nightclub licence in Sydney's hotspot of Kings Cross in Sydney's inner-east costing AUD$60,000 (US$38,000), to restaurants varying between AUD$500 (US$300) and AUD$15,500 (US$9,800). As a point of comparison,

Victoria, under its 1998 Act, had a single licensing application fee of AUD$585 (US$370), with an annual renewal of AUD$175 (US$110).

Moore's Bill advocated amending existing legislation to allow restaurants to supply alcohol without a meal, provided that food service remained the predominant activity, and tables and chairs were available to at least 70 per cent of patrons at all times. It also proposed the introduction of a new small-bar category of licence for bars with 120 patrons or fewer, with only a AUD$500 licence fee (Moore 2007). The amendment, it was argued, would 'promote a more civilized mix with smaller venues that reduce anti-social behaviour by making it easier for staff to know when someone has had too many'. The amendment would also 'encourage small vibrant boutique bars and places to listen to live music, providing an alternative to massive TV and poker machine dominated beer-barns filled with binge drinkers' (Moore 2008). An online support group known as Raise the Bar was formed and the proposals received support from Labor, Liberal, Green and Independent members of parliament (www.raisethebar.org.au).

Moore's Bill was not actually successful. Rather than amend the existing 1982 Liquor Act as Moore had proposed, the NSW government formulated an entirely new Act, which drew upon a series of measures and reforms concerning competition and harm minimization. The Act that came into affect in NSW in July 2008 fundamentally changed the process of licensing in the state. The Act's rationale was comparable to that in England, Wales and Scotland, in that it sought to do away with a complex and bureaucratic regime. The new Act also allowed licences for small bars and live music venues to be set at a fixed rate of AUD$500 (US$316), and restaurants to serve liquor without a meal for a small fee. Restaurants are still required to serve food and provide for up to 70 per cent seating for patrons.

Not everyone supported the introduction of small bars, be they in the form proposed by Moore's Bill, or the state government's Act. The Australian Hotels Association (AHA) were highly critical and went so far as to say they had received little in return for the millions they had donated to the current Australian Labor Party (Clennell 2007: 4). A particular concern of the AHA was that the change in Victoria's licensing laws had led to a doubling of licensed premises in the state. It is unclear the extent to which this reflected the fact that there were more 'smaller' venues rather than large-capacity premises, but the number of liquor outlets in Victoria certainly increased. Between 1998 and 2006 the number of licensed venues grew by 96 per cent. In the same timeframe, licensed venues in NSW increased by 34 per cent (Moore 2007). That more venues equals more alcohol consumption is not necessarily the case, however, and in evidence presented to the NSW Parliament citing a report conducted by Niewenhuysen, the architect of Victoria's laws, there was no increase in per capita consumption in Victoria over this period (Moore 2007).

Nonetheless, the AHA remained critical of the new legislation, pointing to an increase in late-night disorder in Melbourne.

The AHA were critical for other reasons, both of which require a rudimentary understanding of Australian culture and licensing laws. First, across Australia all persons selling alcohol are required to undertake training in the responsible service of alcohol. This covers all providers, whether full-time or casual, from volunteers serving alcohol at a social or local sporting event, airline stewards and bar-staff, to staff in the off-trade and wine makers conducting a wine-tasting. Failing to comply results in a large fine for both licensee and server. Different qualifications are required for staff working in different types of premises and certificates completed in other states are also not directly transferable, with a requirement for holders to undertake a bridging course before working in another state. Mandatory training for those employed in alcohol-related industries remains contentious and evidence points to a variable rate of success in tackling alcohol misuse (Howard-Pitney *et al.* 1991; Lang *et al.* 1998). The training schemes are also currently under review by the NSW Casino, Liquor and Gaming Control Authority. Nonetheless, for the AHA, 'responsible service' was a point of contention. Small bar owners would still have to undergo the same training as all other staff, but the AHA were concerned that small bar owners would not have the same necessary experience as those who ran large pubs/hotels.

Melbourne and Sydney have also long been rival cities. For supporters of the new Act, the promise of small, intimate bars was not only a draw for tourists, but for residents as well. Australia's former Prime Minister and Sydney resident, Paul Keating, was especially scathing of Sydney's pubs:

> The pub culture in Sydney is stultifying bad. It's raucous and it's noisy in their Klondike-like saloons. All that's missing is Lola Montez. Melbourne has got a level of sophistication Sydney doesn't have. You don't get all this guffawing and noise.
>
> (Creagh 2007)

Other prominent supporters, including celebrity chefs and restaurateurs, joined him in his concern. The debate was subsequently framed as Klondike Sydney pubs versus sophisticated Melbournians, or, for John Thorpe, president of the AHA, 'high' versus a more traditional 'Australian' culture. Thorpe argued: 'We aren't barbarians, but we don't want to sit in a hole and drink chardonnay and read a book.' Thorpe's comment went to the heart of the Melbourne–Sydney rivalry, with Melbournians, as the most European Australians, cast as chardonnay-drinking cultural elites. When Moore's Bill was originally tabled, it drew similar criticism from the Opposition's gambling and racing spokesperson, George Souris, who argued that: 'The concept of trying to introduce some sort of European liquor accord involving coffee shops and food outlets is unwork-able. NSW is not Paris' (Smith 2007a).

As in the UK, the Bill – and the subsequent Act – also occurred against the backdrop of increased violence associated with alcohol. A report from the NSW auditor-general noted that across the state there had been an increase from 10,305 alcohol-related assaults in the year from June 1997 to June 1998, to 20,475 alcohol-related assaults in the 2006–2007 period. This figure did not include domestic incidents. The report also found that in the same ten-year period 'alcohol-related malicious damage' had increased by 87 per cent, while 'offensive conduct' had increased by 70 per cent (NSW Audit Office 2008). In light of these debates, Moore's original Bill was criticized for simply adding more venues. George Souris also claimed that 'alcohol-related problems like binge-drinking, antisocial behaviour, domestic violence and youth alcoholism would potentially be exacerbated by a streetscape lined with alcohol outlets' (Smith 2007a).

Despite these objections, the government's Act passed with support from both the Labor Government and the opposition parties. Small bars are now permitted to open from 7 a.m. to 11 p.m. on Mondays to Thursdays, 7 a.m. to 1 a.m. on Fridays and Saturdays, and 10 a.m. to 11 p.m. on Sundays. Extensions are also available. Known as 'general bars' in the Liquor Act, they are not allowed to provide gaming or off-premises sales. They are required to conduct a community impact assessment (CIS) and, in a further departure from Moore's original Bill which had proposed an upper limit of 120 people, the new law states that 'patron numbers will be up to the business operator and local councils to determine' (NSW Office of Liquor, Gaming and Racing 2007: 5). The Act also 'encourages' live music, as a counter to the gambling machines and televisions which dominate many of Sydney's licensed venues.

The impact of the Act is yet to be determined. Informal discussions with Sydney residents have found little evidence of a major shift in alcohol pro-vision, and the CIS has been criticized as overly bureaucratic. Since the legisla-tion is, at the time of writing, still new, this should not be altogether surprising. Nonetheless, what is more important for the discussion at hand is the manner in which Moore's amendment and the subsequent Liquor Act passed by the NSW Labor Government were framed in terms of a 'European' and 'civilized' form of consumption. While it was Melbourne, rather than a notion of 'contin-ental Europe' that served as the inspiration, the new laws were passed in a lan-guage comparable to the passage of the Licensing Act in England and Wales: continental and civilized venues versus large 'booze barns' full of noise and binge drinking. At the heart of this debate is a presumption that the stereotypi-cal form of continental drinking experience (sitting down and eating) in small, locally operated bars will automatically eradicate alcohol misuse.

Discussion

The examples of Barcelona and Copenhagen undermine a simplistic concept of a self-regulating drinking culture that is a supposed feature of the European cultural heritage. Urban areas of mainland Europe and Australia have experienced similar problems to the UK wherever there has been any unregulated or unforeseen expansion in licensed premises that outstrips the capacity of an area and its regulatory agencies to cope. The negative externalities associated with excessive concentrations of alcohol-related premises are ubiquitous and include noise, disorder, anti-social behaviour, low-level crime, occasional outbursts of violence, traffic congestion and litter.

Yet, European cities compete with each other and with other cities internationally for tourism and inward investment. Clubs, nightclubs, bars and restaurants form a necessary part of the cultural 'offer'. They are intrinsic to the public realm, in a physical and symbolic sense, making up an important part of the ambience and identity of places and spaces. City authorities need to promote and encourage nightlife in addition to regulating it. As has been demonstrated, this is a difficult balancing act in a fast-moving and dynamic sector. It is not surprising that many cities have moved towards de-regulation and liberalization, only to step back once the negative impacts started to outweigh the benefits. This was the situation in Reykjavik in Iceland, where in 1999 the Municipal Council liberalized opening hours. Initially this smoothed out the 'peaks' in concentrations of customers leaving premises at the same time and reduced taxi queues, but the total number of admissions to the hospitals' accident and emergency departments increased. Local residents were also disturbed by the prolongation of nightlife. Difficulties were encountered in street cleaning, meaning tourist families were affronted by the previous night's debris as they set off sight-seeing for the day. Eventually the police and the City Centre Steering Group asked for and got a re-institution of the restrictions on opening in response to growing alcohol-related problems in and around the city centre. Eire too, pulled back on its initial extension of opening hours from 11.30 p.m. to 12.30 a.m. on Thursdays following increases in alcohol-related offences and accident and emergency admissions (Plant and Plant 2005).

The problem for local governance is that once liberalization has occurred, it is impossible to move backwards in time to the situation that pre-existed the loosening of restrictions. While limitations on opening hours can be re-instated, although this has difficulties for the operator's profit margins, and operating conditions tightened, it would take a brave authority to revoke licences and planning permissions. Owners and licensees would immediately sue the authority for loss and business confidence would be damaged. The example of Nyhavn in Copenhagen and Temple Bar in Dublin suggest that reparation of the image and status of an area that has been subject to a saturation of licensed premises can be effected through good management practice. The full impact of the new Act in Sydney, Australia is, at this stage, difficult to determine.

Nonetheless, this chapter has highlighted two key arguments in the 'continental model'. First, European cities grapple with comparable issues to the UK and Australia in terms of alcohol misuse. The fundamental difference has been a willingness to impose restrictions on expansion and to better manage the impact of late-night cultures. Second, there are limits to simply adopting a foreign drinking model. A point raised in previous chapters, and worth re-iterating here, is that drinking practices and late-night cultures emerge as a result of a complex network of legislative, cultural, economic, political and social trajectories. Context is always important and a simple change of legislation, or merely the aspiration to 'be more European', needs to be balanced by a range of other interventions. As documented here, rather than relying on an assumed 'natural' inclination towards sobriety, European cities have implemented strategies to manage and control their late-night urban areas. Where they have not, as occurred in Barcelona, problems directly comparable to those in the UK have emerged. In common with cities elsewhere, British town and city centres have also been forced to tackle alcohol-related disorder by a range of strategies. It is to these that we now turn.

Chapter 8

Planning and managing the night-time city

Rhetoric and pragmatism

> It is an interesting fact that planners do little for the city as an entertainment
> space.... One could and must suspect that planners do not like to enjoy life
> or to be entertained, or to be seen to like to be entertained, even occasion-
> ally. Most planners, being Calvinist-minded missionaries for social justice and
> equity or militant warriors for sustainable development, have deleted enter-
> tainment from their proactive agendas.
>
> (Kunzmann 2004: 389)

Introduction

The scale and pace of the expansion of nightlife in Britain was unanticipated by
local authorities and the police. Responses to it were at first piecemeal, with individual authorities and police forces attempting different measures. As the scale
and severity of the negative impacts of expansion became more widely recognized, central government started to play a role in the gathering and dissemina-
tion of good practice. Other agencies such as the Association of Public Service
Excellence (APSE) and the British Urban Regeneration Association (BURA) also
played a part in holding seminars and meetings. The publication of the *Alcohol
Harm Reduction Strategy for England* (Cabinet Office 2004) by central government started to gather together good practice guidance for the first time. An
update to the *Strategy* was produced in June 2007 (HM Government 2007),

which paid more attention to health issues and to the problems of youth (18–24-year-olds) drinking.

A series of different studies have contributed to a growing body of evidence about practical approaches to positive planning, management and 'damage limitation'. The Civic Trust, which is an independent charity part-funded by the government, brought together this thinking in the publication of its report *NightVisions* (Davies and Mummery 2006), which was based on a series of research projects, three of which were undertaken by the authors of this book. *NightVisions* codified an approach to managing the night-time economy in a ten-point plan. This plan has since been developed into a 'per-formance standard' or 'purple flag' for authorities to demonstrate good practice in the management of the night-time economy within their jurisdiction. At the time of writing, the UK government has included 'purple flag' status in its 'Beacon' awards scheme for successful local authorities.

In an earlier chapter, this book set out the urban renaissance vision for the city at night. This has an image of clean attractive streets, populated by happy people drawn from all ages, ethnicities and income groups, flowing from outside event to inside café-bar, restaurant or cultural event, enjoying a conviv-ial sociality that transgresses conventional boundaries of polite interaction, but nevertheless feels safe and life enhancing. By contrast, an image propagated by the UK media is of dark night-time streets dominated by feral young people engaged in an orgiastic display of lewdity and profanity that is about to explode in savage violence and irrational vandalism. Of course, these two sets of images can actually occur in the same place in the course of one night.

The conditions for disorder and violence have already been set out in previous chapters. To reiterate, these are an over-concentration of youth (16–24-year-olds) oriented venues that are managed irresponsibly, turn a blind eye to or encourage binge drinking, employ ill-trained security staff, serve under-age customers and are overcrowded and hot. On the street, a lack of services and facilities associated with the day-time city lead to problems of street fouling (urination and defecation), spreading litter and violent affray caused by competition and queues for fast-food and transportation. Proximity to housing and hotels also causes problems with disturbance and noise. The worst overall outcome for city managers is for a perception to arise of the city centre, or part of the city centre, as a no-go area. Not only will quieter night-time businesses be unwilling to locate there, but in severe cases the attractive-ness of the town or city centre will fall and established residential populations will leave (Allen and Blandy 2004).

Theoretical framing

The night-time economy poses many challenges to the different tiers and agencies involved in governance. Previous chapters have highlighted the drive in the reform of UK licensing legislation to provide the alcohol and entertainment industry with more freedom over the conditions of the sale of alcohol in terms of hours, functions, location and size of premises. As academic commentators point out (Hadfield 2007; Hobbs *et al.* 2005; Room 2004), problems within the night-time economy are conceptualized by the Labour government in terms of 'responsibility'. Responsibility is to be exercised by individuals, so it is the 'irresponsible' licensee for the individual premises who is to be targeted along with the 'minority' of troublemakers who, to paraphrase a minister, spoil the experience of going out for the rest of us. The underlying tenets of economic growth at any cost are, it seems, left unchallenged. The solutions to violence and disorder are seen as lying in the restraint of problematic individuals and operators, thereby justifying ever-more severe penalties and controls. The most recent phase in the government's national alcohol strategy calls for 'sharpened criminal justice for drunken behaviour ... to bear down on those committing crime and antisocial behaviour when drunk' (HM Government 2007: 6). Characterizing such behaviour as an individual problem rather than a group ritual, the strategy proposes to offer the offender 'advice, support and treatment' where appropriate. Meanwhile, from this perspective, 'partnerships', voluntary agreements and education are deployed to promote 'respectable' styles of consumption (Bannister *et al.* 2006).

It is difficult to refute this analysis. At the time of writing, the latest report to reach the news headlines demonstrates unequivocally that licensees in seven different city centres across England were 'routinely guilty' of irresponsible practices such as serving under-age drinkers and those already drunk (KPMG llp 2008). Central government's unwillingness to do more than encourage voluntary codes of responsibility on operators exposes their acceptance of the status quo of 'determined drunkenness'. Even so, it would be untrue to suggest that this acceptance could be read as an outright encouragement to unmediated consumption. The series of changes to the *Guidance* to the Licensing Act 2003 have demonstrated that central government does yield to external pressure. Furthermore, the critics that have argued that local councils were bent on policies to woo the major pubcos and other operators into their town centres (Chatterton and Hollands 2002; Hadfield 2006), are drawing on an evidence base that reaches back to the policies of the mid-1990s. This was a period when the UK planning system was directed by a neo-liberal ideology that sought to reduce and restrict its role (Thornley 1991).

Since 1997, the British state has experienced a conundrum as regards local economic development. On the one hand, as Room (2004) points

out, it is as friendly to big business as the Conservative administration that pre-
ceded it. On the other hand, it has reasserted the significance of spatial planning
and the primacy of place in the Planning and Compulsory Purchase Act 2004.
This Act reasserts the values of social democracy, with its emphasis on local consultation and the need to consider the social and environmental impacts of development. The promotion of 'partnerships' between local public and private sector
stakeholders can be explained as an attempt to adapt to mediate in a mixed
market economy in which economic goals have to be reconciled with the needs
and aspirations of civil society. Hobbs *et al.* (2005) criticize local authorities and
government agencies for entering partnerships, arguing that they assist the
growth of the night-time economy in its present form in the UK, a form that they
characterize as 'an imperfectly regulated zone of quasi-liminality awash on a sea
of alcohol' (Hobbs *et al.* 2005: 165). While this may be true in terms of an overview, it nevertheless overlooks some areas of complexity and contradiction.

To elaborate further, once determined, a permission to develop
cannot be revoked (Tiesdell and Slater 2006). It is therefore entirely reasonable
for local government to attempt to ameliorate and mitigate the unforeseen
negative impacts of past decisions. Furthermore, local government in Britain,
while subject to an increasingly centralized framework of financing and guidance, does enjoy a certain degree of autonomy. This applied particularly to the
circumstances in which Britain's night-time economy expanded. One of the
paradoxes of licensing reform is that it was one of Britain's leading Conservative local authorities, Westminster City Council, that was most vocal in its
opposition to what it saw as a laissez-faire and de-regulatory measure. Local
governments can therefore act in opposition to central government. In the case
of the night-time economy, they could, and did take a lead in promoting new
forms of governance and control.

A further complexity lies in the contested arena of public space. Not
only is this the finite physical space in which many of the manifestations of the
night-time economy are experienced, both positive and negative, but it lies at
the nexus of a complicated series of interrelationships between the social and
the imaginary, two of the three components of Lefebvre's triad (Lefebvre
1991). Much of the discussion of 'liminality' refers to the potentially trans-
formative qualities of communality in public space, qualities that the night-time
economy promises but most frequently fails to deliver (Grazian 2007). Public
space, as Worpole and Knox (2007) explain, is 'co-produced' through its physical configuration, the activities that surround it, the people who inhabit it at any
point in time and the regulatory regimes that control and manage it. Although
central and local government policies do have an impact on the production of
public space, theirs is an influence and not a determinant.

There are a plethora of agencies who are implicated in this 'co-produc-
tion', such that a spreadsheet or flow chart that sets out to describe each rapidly

becomes highly elaborate (Carmona 2004). Different departments in local government, the police, health authorities, local businesses, catchment populations, regional agencies and authorities such as public transport providers, are each implicated. Public space is also an arena for professional dispute and disengagement within recent history, with architects focusing on the building-as-object, highway engineers on traffic management and landscape architects on restricted areas of green and hard landscaping (Marshall 2005). It is only in the last decade that the public realm has risen in importance as a subject for study, and in the last five years as a focus for specific professional practice in the UK. As will be demonstrated in the discussion that follows, responsibility for what happens in the public realm is frequently blurred, or passed from agency to agency.

This discussion elaborates on the complexity surrounding the management of the night-time economy. This suggests that the critical questions could usefully shift in emphasis from a swingeing critique of the hypocrisy of central government to a more nuanced discussion that recognizes the extent to which local governance has to manoeuvre between business interests, the views of the local electorate, local political agendas and national legislation and regulation (Hadfield and Traynor 2008). In the next section of this chapter we set out the ways the night-time economy has been managed at a local level, largely in response to processes that were set in motion in the mid-1990s. At the end of the chapter, we shall return to longer-term visions for the future and assess the potential for spatial planning. First, we shall look at management from the perspective of the different agencies involved.

Managing the night-time economy: policing

At the turn of the millennium, in major centres such as London's West End, Manchester and Nottingham, the police feared losing control. The Greater Manchester Police were one of the first forces in the UK to recognize the need to address the problems of crime, safety and the night-time economy. The force set up a specialist unit in September 2000 in response to the 242 per cent increase in the numbers of licensed premises in Manchester between 1997 and 1999 and the associated 227 per cent increase in crime over the same period (Home Office 2007a). The initiative had four objectives. These were: to reduce the number of serious assaults and, in particular, glass-related injuries; to work in partnership with the licensed trade to achieve better management of premises; to encourage safer drinking practices; and finally to reduce the perception of Manchester as a drunk and disorderly place at night.

Twenty different initiatives were launched to tackle these objectives. Many of the measures taken were successful and continue to be developed by police forces across the UK. The scheme proved effective and in the financial year 2000/2001, the number of serious assaults in Manchester city centre fell by 8.1 per cent. In 2001/2002, a further 12.3 per cent reduction was achieved (Home Office 2007a). Greater Manchester Police have continued to fund the project because one of the key findings from its inception was that to be effective it relied on being resourced with a dedicated team of officers. In 2005 the project was extended from the city centre across Manchester and was re-named Manchester City Safe (www.citycentresafe.com).

The 20 initiatives were a mixture of schemes that supported central government views and others that challenged them. As was noted in the previous chapter, the Licensing Act 2003 focused on the licensed venue as a source of anti-social behaviour and, outside designated Special Policy Areas, restricted the remit of the Act to a small area surrounding the individual establishment. Official targeting of clubs and bars that have a lax approach to legislation and other sorts of standards has become commonplace, with police officers and licensing officers co-operating to draw up a list of the 'top ten' premises where problems have been most frequently recorded and then adopting a multi-agency approach to inspection and enforcement. A multi-agency visit can include officers from licensing, environmental health or protection, trading standards and the fire service. Since 2004, the police have had powers to close down premises that are causing serious nuisance through noise, at 24 hours' notice. The Licensing Act 2003 also gives powers for immediate closure if there is serious disorder.

The focus on responsibility continues with encouraging licensees and managers to improve the management of their premises. Best Bar None awards were introduced in Manchester in March 2003 with the support of the drinks industry. Premises reaching the right standards displayed a plaque outside to inform and attract customers. The standards included providing safe premises, no under-age serving, good neighbourhood relations and a practice of drug awareness. Since its inception in Manchester, 30 other cities and six London boroughs have taken up the scheme, and the Home Office has now joined it as a national partner. This scheme has its own web site and national presence (www.bbnuk.com).

The Licensing Act 2003 has also enabled licensing authorities to carry forward controls over the ways that licensed premises are designed and operated as conditions of the licence. Typical suggestions for conditions cover a checklist that includes CCTV, arrangements for door supervision, searching for drugs, entry requirements such as proof of age, the prohibition of serving bottles, toughened glasses, restricted drinking areas, capacities, queuing, last admission times, music and dancing arrangements and others (Nottinghamshire Police n/d). Dance clubs are required to adopt the *Safer Clubbing Guidelines* (Webster *et al.* 2002), which sets out advice and standards concerning several matters, including adequate ventilation, the provision of tap water (to combat dehydration) and avoiding overcrowding. This has recently been updated with a new, more comprehensive document titled *Safer Nightlife* (Webster 2008) that covers advice, monitoring and recommendations for increased intervention and support by venue staff.

Premises managers employ their own security staff and specifica-tion of their numbers increasingly forms an element in licensing conditions for larger venues. Door staff are the intermediaries between security inside the venues and the external spaces (Graham and Homel 2008). Prior to the intro-duction of the Private Security Industry Act 2001 and the establishment of the Security Industry Authority, door staff acted in effect as unregulated private police (Hobbs *et al.* 2003; Hobbs *et al.* 2000). Their remit extends to the imme-diate vicinity of the premises, monitoring the queue and the immediate curti-lage. Although the training that is now required for door staff or 'bouncers' should have eliminated the abuses recorded prior to regulation, respondents in our focus groups in 2005 were still complaining of irrational aggression on the part of 'bouncers' and their 'winding up' of patrons by making arbitrary decisions about entry. These types of provocative behaviours should, however, be reduced as the training programmes become more widely diffused. In response to the numbers of women coming out in the evening, mainstream clubs and bars are now employing female door staff, and indeed the gender of the door staff is occasionally a requirement of the licensing conditions. A recent study found that female door staff were as willing as their male counterparts to enter violent confrontation if need be (Hobbs *et al.* 2007).

The Manchester City Safe project introduced a 'Nite Net' radio system linking door staff in pub and clubs, CCTV operators, street cleaners, town hall licensing officers, parking wardens and other night-time workers together with police officers in the night-time hours. They can then warn each other of aggressive customers who have been ejected from venues and can target police activity towards particular groups of troublemakers. This initiative was an extension of local 'Pubwatch' systems whereby licensees and mini-cabs used radio and CCTV links to warn each other of troublemakers circuiting from one venue to another. 'Pubwatch' schemes are now in general use across the UK and 'Pubwatch' has had its own website and national organization since 1998 (www.nationalpubwatch.org.uk). The Licensing Act 2003 recommends that all licensees join their local 'watch' scheme. Although supported by the government, these schemes are voluntary.

The use of CCTV during the day-time has burgeoned in recent years in the UK, with many major city and town centres in Britain now having soph-isticated surveillance systems. Television producers in the UK have mined these for entertainment, as Hayward and Hobbs (2007) have noted. Pro-grammes use CCTV footage to show violent incidents, interspersed with shots from the control room and commentary from the police. UK viewers will have gleaned from these a working knowledge of police practice and procedures, from report to arrest. For readers outside the UK who may be less familiar with this technology, CCTV surveillance systems typically have a control room in one part of the city centre, staffed in shifts. Cameras located on the streets feed images to screens in the control room. Some cities have mobile vans, with cameras connected so they can follow trouble. The funding for these control centres comes from a combination of sources, frequently Home Office grants for the equipment, with the revenue funding supplied by the police, the local authority and occasionally a regeneration partnership that includes major businesses.

Although the efficacy of CCTV is contested, a Home Office-sponsored research study found that, when used for alcohol-related disorder or drugs offences, both connected to the night-time economy, it was useful as a deterrent and as a means of identifying offenders. Surveillance offered the police a targeted form of crowd control, enabling them to move on large gatherings of individuals. More contentiously (Coleman 2004), the report noted that CCTV assisted in 'moving on what many operators termed "undesirables", such as beggars and on-street traders' (Gill and Spriggs 2005: 117). CCTV in the night-time economy is therefore used as a way of reducing the degree and extent of violent and potentially explosive situations. There is some evidence that this is manifested in the degree of injury experienced by victims. A comparative study of night-time admissions to an accident and emergency unit in south Wales, from a town centre without CCTV and one with, found that victims from the surveyed town centre experienced less severe injuries (Sivarajasingam et al. 2003). Many studies of CCTV (Griffiths 2003) have repeated the finding that it is not a panacea in itself with regard to crime reduction. The placing of cameras, town centre lighting, procedures for storing and making use of the multitude of images produced, the training of operatives and the information supplied to the public all influence its efficacy (Welsh and Farrington 2002). Furthermore, evidence of frictions and difficulties in the relations between the different agencies and individuals involved suggests that reliance on CCTV as a 'technological fix' is misplaced (Smith 2008).

The voluntary nature of the initiatives so far described form a contrast to the increased powers that central government has given to the police to control alcohol-related crime. The Parliamentary acts that awarded these new powers cover a wide range of offences. One particular concept that lies behind the powers, anti-social behaviour, has been criticized for blurring the boundaries between civil and criminal offences and effectively criminalizing large sections of the population, particularly young people (Gask 2004). While a convincing argument can be made that an inclusive city requires some tolerance of difference and a certain degree of 'incivility' (Bannister et al. 2006;

Whyte 1980), these arguments generally refer to isolated street drinkers rather than multiple subjects who are grossly inebriated and unpredictably threatening. In assessing the measures now available to the police with regard to alcohol-related crime and disorder, there is a difficult triangulation to be made between three poles. These are: the demonstrable evidence of the real threat that alcohol-related crime and disorder brings; the threats to the human rights of those brought into the criminal justice system for low-level incivilities such as shouting or public urination; and finally the efficacy of any measures brought in. In the contested space of the night-time economy, it cannot be assumed that because a power exists, it is enforced, or that it is effective in the reduction of violence.

The City Safe project piloted the successful designation of zones, or designated public places, where the public are not allowed to drink alcohol from open cans or bottles, unless they are in a defined space outside a café or bar. Prior to this, local authorities had been able to pass by-laws banning the open drinking of alcohol, and indeed 42 had done so. However, an independent study concluded that while the by-law had had a positive effect, especially in relation to rowdy and anti-social behaviour, it had not had any effect on under-age drinking or other types of criminal activity (Working Solutions 1998). The Manchester pilot found that rigorous enforcement by highly recognizable and uniformed police officers was critical. Officers had been reluctant to use their powers because there was nowhere for cans and bottles to be thrown away once they had been seized. Specialist bottle and can banks were consequently installed adjacent to bus stops and taxi ranks (Civic Trust 2005). Subsequently, the Criminal Justice and Police Act 2001 gave all local authorities powers to make 'designated public place orders'. These have been widely taken up across England and Wales. The most recent controversial extension of this designation has been to the whole of London's public transport system, when on the eve of the ban a 'spontaneous party' was organized on three Underground lines, leading to the closure of a major national rail station for part of the evening, along with several arrests.

People who drink in a designated public place, who urinate in public or are generally drunk and disorderly can be issued with a penalty notice for disorder (PND) by a police officer without having to be taken into custody, under the powers of the Criminal Justice and Police Act 2001. The object of this is to increase the amount of time that the police spend on the streets and reduce bureaucracy. Under the Violent Crime Reduction Act 2006 these powers are extended further and a drinking banning order can be issued to ban an individual from a specified area for up to 48 hours. This replaces a previous power that enabled licensees to ban individual customers. The police also have powers to disperse groups of people from an area under the Anti-Social Behaviour Act 2003. This power, whose intention was primarily directed at breaking up gangs of threatening youths in residential areas, does not carry the stigma of a criminal offence.

The one area in which central government has been prepared to adopt a tougher stance with the drinks industry has been in the serving of alcohol to under-age drinkers. Three national Alcohol Misuse Enforcement campaigns have used 'sting' operations to test whether licensees or bar-staff are serving under-age drinkers. The first of these found widespread flouting of the law (HM Government 2007). The Violent Crime Reduction Act 2006 gives further powers to the police to close down premises for 48 hours that are found to be making sales to under-age drinkers, and to institute proceedings for the revocation of a licence if this has occurred more than three times in three months. The Licensing Act 2003 increased the maximum fine to £5,000. Nevertheless, a review commissioned by the UK Home Office found these laws were 'frequently' disregarded and people under 18 years old were allowed into venues where they were meant to be prohibited (KPMG llp 2008). Other good practice guidance was ignored, such as DJs encouraging people to drink more and to down spirits, 'shots' and 'shooters' in one draught.

The application of these different powers demands a higher level of police presence in the evening and at night. The policing of public spaces at night has been extended through lower-cost additions to the 'policing family'. There are two categories of officials, police community support officers (PCSOs) and local authority wardens. PCSOs report to the police and are jointly funded by the police and other private or public sector bodies. They have various powers conferred on them by the Police Reform Act 2002 and subsequent statutory orders. These include the ability to confiscate alcohol in designated public places and to stop littering. Local authority wardens report to the local authority and are often funded through regeneration partnerships. Their powers are less than the PCSOs and their role is to act as the 'eyes and ears' for the police – they are linked up to the local police force and can call on them for a speedy response. The origins for these new categories of 'police' lay in the Labour government's drive for 'sustainable communities', whereby deprived neighbourhoods and failing town centres would be made safer for residents and businesses. This means that PCSOs and neighbourhood wardens tend to be deployed during the day, rather than in the late evening or early hours of the morning. For example, even in London's West End, the community wardens only work until 10 p.m. at night (Roberts 2004).

The issue of resources appears even more starkly with regard to the other agencies that are 'stakeholders' in the management of the night-time economy. Whereas the new Labour government was elected on a promise to be 'tough on crime and on the causes of crime' and could therefore divert expenditure towards policing and crime prevention, it has less remit with regard to other areas of regulatory activity. Nevertheless, new initiatives have been taken.

Local authorities

Outside of planning and licensing, there are a number of ways in which a local council in the UK can intervene in the night-time economy. With regard to noise, licensed premises are subject to controls in design and operation. The Building Regulations set standards for noise emission from licensed premises to other premises and the outside. Here, the standards for the UK are less than those in mainland Europe (see Chapter 7). The UK government has recently given local authorities more powers over the operation of licensed premises in the Clean Neighbourhoods and Environment Act 2005, such that they can issue a fixed penalty notice with a fine of £500 on premises that are causing a noise nuisance. Much of the noise from licensed premises comes when customers are leaving late at night (MCM Research 2003). A 'responsibilization' strategy is used to deal with this, through encouraging licensees to include a 'chilling out' hour or half-an-hour hour between the final serving of drinks and the closing down of the entertainment and actual closure for the night.

Local authorities have a duty to clear litter from the streets. Late-night takeaways cause particular problems because drunken people tend to drop litter indiscriminately. This is not only unsightly but also, drawing on the 'broken windows' thesis (Kelling and Wilson 1982), contributes to an impression of disorder and lawlessness. Local councils have powers under the Clean Neighbourhoods and Environment Act 2005 to require particular businesses to clear litter. Fast-food outlets are also encouraged to sign up to a voluntary code to clear the litter dropped by their customers (ODPM 2005c). Some authorities have re-scheduled their street-cleaning programmes so that they operate after the late-night venues have closed and before the first commuters arrive in the morning. In Manchester, the street-cleaning teams go into action as the clubs close and the cleaning machines – with their water jets – are used as a dispersal mechanism (House of Commons 2004/5).

Street cleaning has to deal with other by-products of excessive alcohol consumption such as urination, vomiting and defecation. Given the involuntary nature of these bodily functions, criminalizing public urination does not really help in preventing it, and local authorities have been progressively closing public toilets even in the day-time (Greed 2003). To combat the problem, Reading Borough Council introduced 'pop-up' urinals which emerge or are placed on the pavement on weekend evenings. They were shortly followed in this by other authorities. Although criticized by feminists for privileging male night-time needs over, say, the day-time requirements of women, the benefits to local authorities are demonstrated by Westminster City Council's figures for collection. Westminster City Council brings 12 mobile urinals into the West End on Friday and Saturday nights, has two 'telescopic' toilets that are hidden under the pavement during the day and 'pop-up' at night and two

169

fixed urinals (City of Westminster 2008). A preliminary report on the first stage of implementation of these facilities noted that 154 gallons of urine had been collected from 11 urinals in a two-day period, with 4,185 people using them (Elvins and Hadfield 2003: 26).

Local authorities have control over some of the activities that happen on the pavement. Pavement cafés have come to be associated with gentrified city centres that are oriented towards middle-class consumption (Atkinson 2003). Nevertheless, drinking and eating outside is as popular outside a large 'drinking den' as an upmarket restaurant. This has become more marked following the ban on smoking in public places, first enacted in 2007. When Westminster City Council experimented with al fresco dining in Leicester Square, in association with other measures, street crime fell by 70 per cent in 2002–2003 when compared to 2000–2001 (Westminster City Council 2003). The council does have to exert controls over the design and operation of the outside areas in order to reduce noise and to prevent chairs being used as weapons.

Lighting falls within a local authority's remit. Its contribution to crime reduction and increasing feelings of safety is well documented for the wider city and in public space (Cozens *et al.* 2003; Painter and Farrington 2001; Welsh and Farrington 2006). Improving lighting forms part of an integrated approach to improving the management of the night-time economy. The Civic Trust uses lighting schemes as part of its improvement strategies in its *NightVisions* programme and, of course, high levels of lighting are needed to operate CCTV systems effectively.

In addition to these 'front-line' services, many councils in the UK now have economic development units and some also have dedicated tourism support services. In some instances these units use what influence they have to diversify the night-time economy 'offerings'. In Norwich, for example, the City Council has published a brochure outlining the different activities available at night. In Leeds, the city centre management group runs a programme of evening events that have the intention of encouraging families into the city centre in the evening. This programme draws on the ideas first expounded by Bianchini (1995) with reference to Rome (see Chapter 2). Attempts at diversification are limited, however, by local authority budgets and the restricted appeal of cultural facilities (Tallon *et al.* 2006). London presents a different case, partly because travel distances mean that many workers stay on to socialize after work, and because its size and attractions mean that national cultural institutions such as the Tate galleries and the Victoria and Albert museum can afford to stay open later on weekend evenings and run a programme of evening events. Outside of London, even persuading retailers to keep their shops open later than 5.30 p.m. on one night of the week poses difficulties for town centre management partnerships.

Transport providers

Taxis and buses are particularly important for the night-time economy. In the UK these services are run by private operators, under the control of the relevant transport authority in the case of buses, and as licensed by the local authority for taxis. The expansion of the night-time economy would, by its very nature, appear to support a growth in both types of service. Certainly in London, where from 2000–2008 the mayor provided substantial political leadership, the night bus network increased from 57 routes to over 100, some of which are 24-hour services. All operate a seven-days-per-week service (GLA 2008). Successful night bus services have been established in other cities with a buoyant day-time economy and a thriving growth in bus demand, such as Manchester, Bristol and Brighton. In other towns and cities, such as Bolton, Ipswich, Folkestone, Leicester and Colchester, new services have been provided, but have not yet proved to be commercially viable.

Taxis in Britain fall into two categories. Vehicles with meters, normally black cabs, can be hailed from the street. 'Mini-cabs' or private hire vehicles are prohibited from taking fares from the street and have to be pre-booked. Local authorities, or, in London, Transport for London, regulate both types. The Department for Transport has encouraged local authorities to issue more licences to meet late-night demand. Although local authorities can issue licences they cannot compel drivers to meet demand, and problems remain in many locations with long queues for late-night taxis.

Security is an issue in late-night transport, for both drivers and customers. The entire fleet of London buses now has CCTV installed to prevent attacks between passengers. Mini-cab and taxi drivers are frequently deterred by the possibility of verbal abuse or physical assault. Radio links between the cabs and the police help to provide a measure of safety and are becoming more widely adopted. In some cases unlicensed taxis have proved to be a danger for lone women, where they are vulnerable to sexual assault and rape from male drivers with a history of criminality. Transport for London ran a successful campaign over a number of years 'Know What You are Getting Into' and have promoted schemes to provide easy access to the phone numbers of licensed taxis and mini-cabs. This has led to a 46 per cent decrease in the number of sexual attacks in 2007 compared to 2002 (Mayor of London 2007).

Queues for taxis are flashpoints for violence late at night. Transport for London, in association with the police and the London boroughs, has established several taxi ranks across London that have 'taxi marshals' to regulate the queue. This scheme was used in the 1990s in Times Square in New York and has been successfully transferred to the UK. The scheme in Kingston, a town centre on the fringe of Greater London, is funded by the mini-cab operators, but employs accredited door staff from the local nightclubs (GLA 2007).

171

Centres outside London have experienced more difficulties in funding such schemes, but many experiments have been initiated both in major centres such as Liverpool and Wolverhampton and in smaller towns such as Newcastle-under-Lyme, Mansfield and Ayr.

In the management of the night-time economy, the key players are the police, local authorities and transport regulators and providers. Other agencies make a lesser, but nevertheless significant contribution. Local accident and emergency centres receive the casualties of the late-night economy. Their records, together with ambulance call-out rates, provide an invaluable database for the recording of alcohol-related injuries and illnesses (Elvins and Hadfield 2003; Town Centres Ltd 2001). Accident and emergency departments in Liverpool have shared information with Citysafe, its community safety partnership, so hotspot locations and bars can be targeted (HM Government 2007). Cardiff was one of the first health trusts to provide a mobile triage unit in its city centre. This was implemented as too many of its ambulances were being deployed late at night on weekends and were unavailable for other serious incidents in the wider area (Metropolitan Police Clubs and Vice Operational Command Unit 2003). This practice has been taken up in other towns and cities. Health workers are now being encouraged to be more pro-active with patients presenting with alcohol problems, and to offer advice and support.

Health workers have made interventions into drug use in nightclubs and 'party' venues. Drugs and alcohol action teams provide advice and support to event promoters and licensees, in addition to outreach information and advice to clubbers. Other agencies, such as Crew 2000 in Edinburgh, monitor and evaluate drug use, thereby providing useful information to agencies so that they can answer needs and anticipate new developments. These agencies have similar counterparts in mainland Europe (Energy Control 2007). These initiatives, which recognize the prevalence of illegal use of drugs and admit that attempts to totally eradicate it are unrealistic, sit in contradiction to UK government pronouncements on its intentions towards 'sharpening' the use of the criminal justice system with regard to nightlife.

Another statutory agency, the Fire and Escape Service, may seem an unlikely stakeholder in nightlife. Fire presents a substantial risk in all types of entertainment venues and the presence of large numbers of people under the influence of alcohol and drugs exacerbates the threat. The Fire Service is a responsible authority as regards licensing applications and has to be consulted. Overcrowding and capacity issues are important, as are the operating schedule, design and arrangements for fire prevention. Fire officers may get involved in multi-agency enforcement visits.

Non-governmental organisations such as the Red Cross and the St John's Ambulance Brigade frequently attend special events such as festivals. Student unions are probably the largest of the voluntary organizations involved in nightlife. The most widely known intervention by voluntary agencies, the police and local nightclubs in the UK is the SOS Bus in Norwich (see Davies and Mummery (2006) for a full description of the project). The 'bus' provides a temporary shelter for 'abandoned incapables', that is young clubbers who have become isolated from their friends and find themselves without the means or the bodily capability to get home. Outreach workers find them, as well as providing a walk-in service. Workers on the bus offer practical help, advice and referral to other agencies.

The governance of nightlife is therefore not a simple matter of pitching 'producers' against 'regulators' (Chatterton and Hollands 2002). The rhetoric of government urges self-regulation on the producers. Individual producers participate in management, through providing door staff for taxi-rank marshal schemes or funding for rescue projects. The aggressive statements made by central government about the importance of eradicating crime segue into a pragmatic acceptance of illegal drug use at the local level. To square the circle, blame is placed on a 'minority' of licensees and promoters and a small number of offenders. These complex messages and the contested nature of public space lend themselves to a 'partnership' approach on the part of the police and local authorities who, by themselves, lack the resources and powers to control the streets.

Partnership

The first licensing forum bringing together licensees and promoters, the police, licensing officers, environmental protection officers, taxi and mini-cab operators and representatives of the emergency services was established by the Greater Manchester Police. As the success of this initiative became more widely known, other cities such as Nottingham and Leeds followed suit. By 2004, multi-agency, public–private partnerships to tackle night-time economy issues became part of central government policy, endorsed in the *Alcohol Harm Reduction Strategy* (Cabinet Office 2004). These partnerships mirrored town centre management partnerships. In Bolton, the Townsafe Partnership Group was an extension of the Bolton Town Centre Company, the day-time town centre management partnership. Initial funding for a night-time economy manager was provided by the community safety partnership (House of Commons 2004/2005). The types of activities funded by the partnership are typical of many night-time economy action plans and included targeted policing, a commitment to socially responsible practices on the part of licensees, agreements to share information between parties, extending the hours of the CCTV coverage, appointing street wardens or guardians, targeting and coordinating enforcement and running education campaigns in the local media.

Such partnerships have demonstrated their value in reducing certain categories of crime, for example the Liverpool Citysafe partnership work reduced assaults, robbery and anti-social behaviour by 28 per cent between 2005 and 2006 in the city centre (HM Government 2007). They also provide a vehicle for the collection of a robust evidence base about the night-time economy in a particular locality. For example, a planning officer in the London borough of Camden was funded by the Greater London Authority to produce a comprehensive report on the state of its nightlife that contains a wealth of information on which to base a policy response (Mayor of London 2004).

Despite these initiatives, there are difficulties associated with night-time economy partnerships. The UK government took office in 1997 with a commitment to 'joined up' government. With this in mind it promoted interdepartmental working across different tiers of governance and, drawing on a neo-liberal agenda derived from the previous Conservative administration, welcomed and expanded public–private partnerships. The result has been a plethora of partnerships, each of which has its own strategy. Some partnerships do not necessarily coincide with local authority boundaries. Crime and disorder reduction partnerships, for example, are aligned with police authority boundaries. Each partnership strategy has to sit alongside other strategies produced by the local authority and within its corporate plan. The flow chart for Cheltenham's night-time economy strategy, which describes itself as 'the largest clubbing centre between Birmingham and Bristol', is part of a complicated matrix (Cheltenham Borough Council 2004: 3).

Another issue is that of support, by both public and private sectors. In a study by the authors of entertainment operators and licensees in 2003/2004, mixed views were expressed about partnerships. One senior

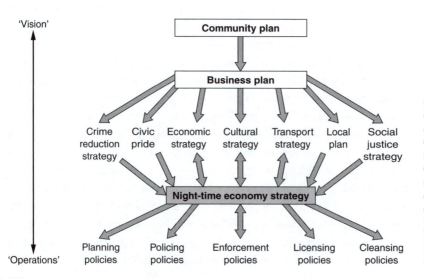

'Vision'

Community plan

Business plan

Crime reduction strategy | Civic pride | Economic strategy | Cultural strategy | Transport strategy | Local plan | Social justice strategy

Night-time economy strategy

Planning policies | Policing policies | Enforcement policies | Licensing policies | Cleansing policies

'Operations'

8.1
Organizational chart for Cheltenham Borough Council's night-time economy strategy.
Source: Cheltenham Borough Council (2004:3), Crown Copyright, 2004.

representative of a major nightclub chain was of the view that meetings that had specified goals were worthy of support, but there were frustrations:

> One of the problems is that, we're a business, we're in things for commercial reasons. When one is getting involved in those types of schemes they tend to become incredibly cumbersome, and very political and bureaucratic, so it's very different to the way that we operate in our day-to-day business and it can be very frustrating and sometimes you really don't see results, and things move very, very slowly, and of course you end up again with people with different interests.
>
> (Senior manager, nightclub company)

Another complaint was about the industry being regarded as a unity, regardless of each company's degree of responsible practice:

> From our point of view, one of our problems, one of our concerns is being tarred with the same brush as everybody else, there's a bit of a sort of conundrum for people in the industry, because when you want to get some change from local authorities or government you need to try and unite and speak with one voice but then that means that you are often getting into bed with people that you don't really want to be associated with. So we want to be identified for what we are and if we don't match up then obviously we say well fine we deserve whatever we get, but we don't want to end up being tarred with the same brush as other people just because we sell drinks.
>
> (Senior executive, pubco)

A further complaint from local licensees was about the turnover of police and local authority staff attending the meetings. The 'churn' of staff moving on meant that agreements and understandings had to be continually re-forged.

When the meetings were well structured and goals were achieved, private sector partners did feel that they were worthwhile to them in a business sense:

> [Our] police in particular are very good, very proactive, and very supportive of the licensed trade and the night-time economy. The towns that do that seem to prosper a lot more. They seem to have less troubles, their A&Es aren't full every Saturday night, people tend to go out and have a good night and a laugh and they know if there is any trouble they could get a) locked up or b) there's someone there to help them.
>
> (Local licensee, ex-mining town)

In Edinburgh, anti-social behaviour is not, according to a police spokesperson, a significant problem. They have, nonetheless, initiated a number of schemes in an attempt to reduce alcohol consumption, and to tackle late-night disorder. In the report *Changing Scotland's Relationship With Alcohol*, produced by the Scottish government (Scottish Government 2008), a framework for action was proposed that called for a series of measures to reduce alcohol misuse. These included minimum pricing, restrictions on promotional material, separate check-outs for alcohol sales in off-licenses, and introducing a fee for alcohol retail outlets to contribute towards the costs of alcohol-related problems.

In interviews, headquarter regional managers reported that they encouraged their local managers to attend local partnership meetings. Given the reported laxity with which social responsibility is exercised by some operators within the alcohol industry, it is easy to entertain the suspicion that for some companies at least, membership of a licensing forum or night-time economy partnership has become part of public relations. An overt impression of social responsibility is maintained, good relationships with the local police force and licensing officers are formed and meanwhile, back at the venue, regulations and guidance can be disregarded as commercial imperatives balanced against the prospects of an enforcement visit allow. This is certainly not to imply corruption or collaboration on the part of the authorities, but more of a disparity between public utterances and private practices. In a sense, this is a mirror image of the gap between the government's aspirations for nightlife and lack of intervention.

Funding partnerships

The funding of night-time economy partnerships is an issue. In the larger city centres, local authorities have sufficient revenue and the problems are of such a scale as to merit expenditure. Smaller authorities, such as county towns, face greater difficulties. Other entertainment and leisure providers, such as cinemas and late-night supermarkets take the attitude that alcohol-related problems are of little relevance to them. Many of the Nightsafe and Citysafe initiatives seem to have been funded by the police or community safety partnerships in the first instance. The Department of Communities and Local Government has promoted the concepts of business improvement districts (BIDs) in the UK. These schemes promise to resolve the issue of providing services at night by levying an extra rate on businesses in a specific locality. Business occupiers are balloted and to be successful a BID has to have the agreement of a numerical majority of businesses, as well as a majority in terms of rateable value. The freehold owners and investors of the buildings are not included in the ballot, but may be on the board. This is in contrast to mainstream practice in the USA, whereby owners hope to see the values of their properties rise through their involvement in a BID (Cook 2008).

At the beginning of 2008, over 60 BIDs were in operation in England, Wales and Scotland. Some of their business plans include measures that assist, for example, a Nite Net radio scheme in Ealing and a taxi marshal scheme and improved training for the licensed trade in Blackpool (National UK BIDs Advisory Service 2008). One BID, Nottingham Leisure, is focused entirely on the night-time economy and is the first BID to be solely intended for the licensed trade. The items included in the business plan for the BID are extra services such as improved lighting, taxi marshalling, taxi ranks, improved CCTV, minor public-realm improvements, promotional brochures and a Christmas event. The proposals were derived from the findings of a market research exercise that examined why footfall in Nottingham was declining. The business plan's main 'pitch' was to combat Nottingham city centre's decline in the face of competition from the suburbs and other centres (Nottingham Leisure BID 2007).

As an alternative to the voluntary nature of BIDs, which do not appear to have the capacity to pay for all of the costs of alcohol-related disorder, central government has given local councils the opportunity to declare small geographical areas as alcohol disorder zones (ADZ). This power is awarded under the Violent Crime Reduction Act 2006. At the time of writing, no local authority had used this provision, although a press briefing from a survey commissioned by the Local Government Association suggested over one-third were considering this option (Local Government Association 5 June 2008). The introduction of the power is a belated recognition by government that alcohol-related crime and disorder may be associated with a density of licensed premises, rather than being attributed to individual establishments (UK Parliament n.d.). The declaration of an ADZ can only be made as a last resort after other powers have failed. Local authorities and the police are tasked with gathering an evidence base and consulting widely before designation. Once in place, licensees within the ADZ have to draw up an action plan, the cost of which is paid for by a compulsory levy on the licensed trade. The draft action plan provided in the guidance to the act is drawn from current good practice and hence includes items such as taxi marshalling, extra taxi ranks, community wardens, increased membership of Pubwatch, increases in dedicated police numbers and more attention to socially responsible service through Best Bar None schemes.

The two faces of government policy are now apparent in the BID and ADZ schemes. It is entirely possible that a large town centre could include a night-time economy-oriented BID and an ADZ. Some larger centres in Britain contain one or two streets, or even a part of one street that are 'hot spots' or 'no go' areas after midnight. Other parts of the centre are more multi-purpose, but would nevertheless benefit from a greater attention to security and

8.2
Cartoon, alcohol disorder zone, first published *Observer*, **23 January 2005.**
Source: Robert Thompson Cartoons www.robert thompsoncartoons. com.

dispersal. The dual nature of this response reveals the confusion that lies at the heart of legislative approaches to alcohol consumption. It is also selective in its processes, focusing solely on business and the consumer. As Cook (2008) points out in relation to BIDs, the views of the wider community are not considered in what are fundamental decisions about the shape and form of urban centres. These decisions are normally part of the process of town planning and it is to this that we shall turn to next.

Planning

The placing of responsibility for licensing first within the Home Office and then the Department for Culture, Media and Sport has tended to marginalize nightlife as a planning issue for British central government. National planning guidance considers nightlife as a minor element in town centre development, worthy of a small number of paragraphs in *Planning Policy Statement 6*, a document that is mainly devoted to day-time retailing. Yet nightlife, as has been demonstrated, plays a significant role in economic and cultural development, tourism and destination marketing strategies. It also poses challenges with regards to the compact-city model of development and the notion of an urban renaissance.

The first chapter of this book set out the case for considering night-life as an important component of town planning. Subsequent chapters have demonstrated that the control of alcohol-related leisure activities is necessary and that these activities are not 'self-regulating', neither can they be left to the free market to resolve, nor placed in a convenient category labelled 'culture'. The addictive nature of alcohol and its associations with violence and other forms of anti-social and criminal behaviour mean that planners have a particular responsibility towards these land-uses in terms of spatial planning and urban design. British law places other responsibilities on planners with regard to crime reduction and safety through the provisions of the Crime and Disorder Act 1998 and the Gender Equality Act 2006. The removal of the concept of 'need' in British licensing law now firmly places the notion of what an appropriate number and density of alcohol-related outlets might be for a neighbourhood or town centre within the remit of town and country planning.

Planners have only recently been confronted with the negative impacts of the expansion of nightlife and have had little time to contemplate what might be the appropriate tools for a positive approach towards the night-time city. Policy guidance tends towards aspirational statements rather than practical tools and measures. Current UK national guidance urges local planning authorities to:

> prepare planning policies to help manage the evening and night-time economy in appropriate centres. These policies should encourage a range of complementary evening and night-time economy uses which appeal to a wide range of age and social groups, ensuring that provision is made where appropriate for a range of leisure, cultural and tourism activities such as cinemas, theatres, restaurants, public houses, bars, nightclubs and cafés.
>
> (ODPM 2005b: 12)

While laudable, this guidance does not provide any direction on how to achieve this diversity. Instead, the emphasis lies on ensuring the retail viability of town centres and controlling out-of-town development. This issue will be discussed further in Chapter 9.

A joint ODPM and Home Office document, *Safer Places*, provides advice on reducing alcohol-related violence. The guidance combines the principles of good urban design with the tenets of a *Secured by Design* approach and the 'broken windows' thesis. Planners and urban designers are urged to

179

pay attention to providing clear, 'legible' channels for movement that are structured in an easily accessible and logical way. Populating spaces is important, as
is ensuring that there are 'active frontages' along the key movement routes.
Dark alleyways and passages formed with blank walls on either side are to be
avoided or gated off. The document argues that possibilities for criminal activities are reduced if ambiguities of ownership are removed and public space is
either clearly public or indisputably in private ownership. The report recommends 'target hardening', that is making it more difficult to commit opportunist
burglaries or acts of vandalism. The report moves beyond design to recommend keeping up standards of maintenance and management in public space.
In a similar vein, an integrated managerial approach to the night-time economy
is proposed, so that planning and licensing policies dovetail, transport links
allow the speedy evacuation of customers and a diversification of the evening
'offerings' prevents a concentration of licensed premises in one location
(ODPM 2004b). As a 'good practice' document, *Safer Places* has no statutory
weight and planning officers cannot rely on it to say, refuse planning applications. It does, however, give an idea of what a well-planned and managed
night-time economy might look like.

There are temporal and spatial controls that planning officers can
draw on (Tiesdell and Slater 2006). Planning conditions, for example, can place
demands for extra sound insulation on developers where ground-floor uses
may be noisy. This can extend to the provision of lobbies. Planning conditions
can also place time limits on outside activities, thereby preventing serving outdoors after a particular time or prohibiting the playing of amplified music. Manchester City Council has had extensive experience of placing these conditions
on developers. In addition to planning conditions, development control officers
can demand that developers provide contributions to the local community
arising from the benefits from development, known as 'planning obligations'.
The Royal Borough of Kingston-upon-Thames has been able to use contributions raised through planning obligations to help pay for a second night-time
radio surveillance system (Roberts 2004). Planning obligations suffer from the
limitation that they are negotiated on a site-by-site basis and tend only to be
required for larger developments.

Moving beyond the individual building, planning documents do have
the capability of setting out a long-term framework for the development of an
evening and night-time economy. Some examples of good practice were beginning to emerge as knowledge of the pitfalls of an unrestrained night-time
economy grew. In Sheffield, the supplementary planning guidance for the
Devonshire Quarter, which promoted a café culture but restricted a concentration of late-night licensed premises, won a Royal Town Planning Institute award
(Planning Transport and Highways 2001). Westminster City Council produced a
highly sophisticated draft guidance document for three of the entertainment

areas in London's West End, broken down into 17 micro-districts (City Planning Group 2006). The London Borough of Tower Hamlets prevented excessive competition amongst the Asian restaurateurs of Brick Lane by protecting the retail frontages at its southern end (Shaw and Karmowska 2004). Norwich City Council was able to re-conceptualize a concentration of clubs and pubs on the Prince of Wales Road as a late-night activities zone, thereby offering protection to historic neighbourhoods nearby (City of Norwich 2004).

These examples have been superseded by a change in the planning system, brought in by the Planning and Compulsory Purchase Act 2004. The Act offers the potential for more significant interventions in shaping urban areas at night through the introduction of supplementary planning documents (SPDs) and/or area action plans (AAPs) that form part of a local authority's Local Development Framework. The SPDs are thematic, whereas the AAP is neighbourhood-specific. Once these documents have been drafted, consulted upon and formally adopted, they will carry more weight than the supplementary guidance under the previous system. The AAPs will also be more three-dimensional and fine-grained in their specificity. The problem is that the new 'cascade' of planning documents that make up the Local Development Framework is untested and will take time to be put into place. An AAP, for example, takes 18 months to go through all of its stages of consultation and can only be adopted when the core strategies to which it relates have been agreed (ODPM 2004a).

AAPs can take the form of a masterplan or urban design framework. This is advantageous with regard to night-time issues of noise and disturbance since they describe an area as a three-dimensional entity. When used in combination with design codes, such documents could specify the preferred locations of particular entertainment uses and their relationship to the street and to other uses, such as residential (DCLG 2006a). Research conducted by the authors has found these factors to be important. For example, a study of four entertainment areas in London's West End, Temple Bar in Dublin, Nyhavn in Copenhagen and the Hackescher Höfe in Berlin found that reported problems with noise and disturbance were experienced more acutely in the London and Dublin locations than in Berlin and Copenhagen. While the management regimes differed widely (Roberts *et al.* 2006), the configuration of the street layout varied between locations. In Dublin and London, the streets were narrower and the block size was smaller, thereby 'capturing' sound from the street. In Copenhagen and Berlin, the block sizes were larger and the streets wider.

The depth of the blocks in Copenhagen and Berlin allowed flats to be gated off from noisy thoroughfares where people were likely to pass after a night out. In the new construction in Temple Bar this tactic was used to protect residents' amenity, in both old and new construction. An existing block of social rented housing was fixed with fences and gates around its communal areas.

STREET SECTIONS

8.3
Sections through principal streets in four different micro-entertainment neighbourhoods.
Source: author.

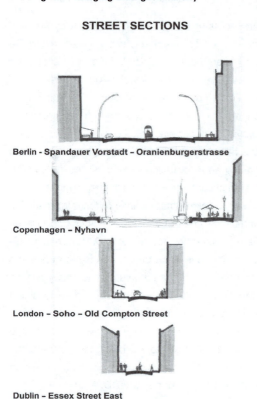

Berlin - Spandauer Vorstadt - Oranienburgerstrasse

Copenhagen - Nyhavn

London - Soho - Old Compton Street

Dublin - Essex Street East

A new 'perimeter' block of flats was constructed around an internal courtyard, gated off from the street. Small shops were provided on the ground floor. These simple provisions, alongside planning and licensing conditions that recognize the need for sleep, enable night-time uses to coexist with housing. The gating off of streets where crime and anti-social behaviour has become a problem can be achieved with a gating order under the Clean Neighbourhoods and Environment Act 2005.

To summarize, planners in England and Wales do have an increased array of powers to control licensed premises. The revised Use Class Orders can frame distinctions between establishments whose primary purpose is to sell alcohol and those whose business is food-based. AAPs can set out guidelines for the development of mixed-use areas and, when aligned with the extra controls offered by Cumulative Impact Areas (Special Policy Areas) under the Licensing Act 2003, can safeguard against over-concentration and protect diversity. However, the drafting of these documents is resource intensive and they also have to be defended in individual court cases against wily licensing and planning lawyers. Further problems may arise in the future when promoters and operators in the licensed trade decide not to challenge regulations in the tightly defined areas of an established city centre or sub-centre, and move instead to a residential suburb. In these situations the opening of a large neighbourhood bar in an unsuitable location can cause as much noise and disturbance as several in a city centre.

Planners also have to be mindful of the dynamic nature of urban culture. Creativity and innovation, as Kunzmann (2004) has remarked, does not necessarily happen in spatially delineated areas. Furthermore, many planners, as he explains in the quotation presented at the beginning of this chapter, do not consider entertainment to be a significant issue for planning. One of our interviewees commented: 'If we had the same approach to the day-time economy ... they'd say we need better car parks, a better bus service, we need to put more people on' (town planner, north-eastern authority).

Those planners who do decide to take entertainment seriously face dilemmas. The preceding paragraphs highlighted the problem of the noisy drinking den disturbing a quiet suburban neighbourhood. The planners' stance can easily be justified in favour of control. Increased regulation is not always the most productive position, however. The emergence of new youth and artistic movements, as was the case with the Hacienda club in Manchester or Cream in Liverpool, can go hand in hand with disturbance, police complaints and illegal drug taking. The extent to which a planning vision should or could accommodate this type of risky innovation is a matter of judgement and political will. The debate about entertainment in Copenhagen's city centre, discussed in Chapter 7, over whether the city should become more 'raucous' or stay 'a little bit small town' typifies these choices. In these situations, public consultation is advisable. We shall turn to the issue of public preferences in the next chapter.

Concluding comments

The management and planning of the environment for the night-time economy in Britain is fraught with confusions and contradictions. Central government policy combines economic boosterism for the drinks industry with increasing regulation and control over individual consumers. Local authorities and the police mediate between these two positions, drawing stakeholder businesses into partnerships in an attempt to provide well-managed town centres with limited resources. The stakeholders themselves are not a unified group and tensions exist between different licensees and between licensees and the statutory agencies. Town planners appear to stand aloof from these processes, historically supporting the opposing demands of economic growth and crime reduction, yet are charged with shaping the overall vision for spaces and places. The resolution of these tensions in managing the night-time economy is forged at the local level and depends on the resources and attitudes of the local stakeholders for its 'success'.

Meanwhile, little is known about the aspirations of those who are excluded from the contemporary night-time economy by virtue of their age, responsibilities for care or a host of other reasons. Debate has focused so determinedly on binge drinking, alcohol misuse and crime and disorder that the fundamental questions raised by licensing reform, about the shape of towns, cities and neighbourhoods, have been virtually ignored. It is to these 'forgotten' consumers that we shall turn to next.

Chapter 9

Consumers

> JEFF: I can't go to the post office to get what they couldn't put through the letterbox ... they should have delivered it in my opinion when I was there and ... they keep posting me these letters through the door box saying 'Sorry, you were out!'
>
> JANE: And they call you back at the same time again tomorrow and you're still not there! [laughs]
>
> (focus group, office town)

> TRACY: If you had somewhere in [office town] which put on a cabaret, have a bite to eat and families were welcome to sit around the table and the kids could have a disco, bit like the clubs but without having to be a club.
>
> (focus group, office town)

Introduction

It was noted in the conclusion to Chapter 8 that 'consumers' represent an absent voice in discussions of town centre management and planning and this can be attributed to several factors. First, as has been argued elsewhere, discussions of late-night culture tend to be dominated by tales that represent the fringes of the late-night city. This is not unusual. As Jackson and Thrift (1995) note, spaces of consumption are typically represented in relation to the bizarre and extraordinary, rather than the mundane. In relation to the city at night, this is especially the case, where, as was discussed in Chapter 2, the themes of fear and pleasure continue to structure the ways the late-night city is understood. This is not to say research into, say, prostitution, raves, drug-use or violence are not important. For many people, these activities are central to how the late-night city is lived, experienced or understood. In other words, despite the disparity and diversity of late-night cultures, the fringe continues to be central to how the city at night is represented. The emphasis on extremes of behaviour is especially the case in terms of drinking, where binge drinking overshadows the more typical patterns of consumption in ordinary venues or at home.

There is a subsequent need to conduct research into late-night culture that focuses on ordinary experiences and leisure spaces. Albeit limited, this work does exist. Moran and Skeggs' research into lesbian, gay, bisexual and transgender communities in Manchester's gay village (Moran and Skeggs 2004); Latham's study of Auckland, New Zealand (Latham 2003); Hubbard's discussion of embodiment and emotion in relation to out-of-town leisure centres (Hubbard 2005); or the significant body of work currently being produced by Valentine *et al.* (Jayne *et al.* 2006; Jayne *et al.* 2008; Valentine *et al.* 2008) represent new trends in thinking about late-night practices and behaviours. Each deals with different aspects of late-night culture in ways that destabilize the traditional focus on the more remarkable aspects of the nocturnal city. This body of work also focuses more on drinking than actual drunkenness and therefore represents a departure from the notion of an homogeneous 'Binge Britain'. Moreover, where drinking practices are placed at the heart of the discussion, this more recent work renders the night-time economy and alcohol consumption as a *productive* activity for the formation and performance of new identities and forms of sociality.

Drawing primarily on the work of anthropologist Mary Douglas (1987), who focused on drinking as a contextually contingent social practice, these more recent discussions of alcohol consumption have examined it in terms of how it functions to enable social relations and new forms of identity formation (Wilson 2005). That is, rather than assuming consumer identities exist prior to the drinking experience, consumption is understood to enable different forms of identity and sociality. To borrow Wilson's example, in traditional accounts of alcohol consumption it may be argued that Irish people drink Guinness and sing songs due to a pre-existing sense of being Irish. A more productive account of consumer practice would argue that individuals, who may or may not be Irish, will drink Guinness and sing songs in order to temporarily perform and feel a sense of Irish identity, with all the associated connotations of community, belonging and 'Irishness'. In this alternative account, consumers are not passive users of cultural products with existing fixed identities, but active users who consume in order to take part in late-night culture in new and possibly unique ways.

This argument extends to the role different venues and events can also play in enabling different forms of consumer identities and behaviours. This trend to examine the productive aspects of the night-time economy, late-night spaces and consumer behaviour can be seen in recent work about the late-night city and race (Talbot 2004) as well as Montemurro's work on bachelorette parties in the USA (Montemurro 2001, 2003, 2005), Kneale's work on nineteenth-century pubs (Kneale 1999), or Lindsay's study of drinking practices in Melbourne's varied 'commercial' or 'niche' pubs and bars (Lindsay 2005).

Shifting the debate away from a pre-determined or universalized model of pleasure versus fear, or debates about crime and health, new trajectories of analysis have therefore been opened that consider a wider array of late-night activities, experiences, cultures and consumers. It contributes to our understanding of late-night cultural practices and forms, as well as demonstrating that there is a good deal more taking place at night than a battle between a frightened responsible majority and 'binge drinking savages' (Eldridge and Roberts 2008). Fear and pleasure, or health and crime, remain central to how late-night town and city centres are understood, but consumers also experience cities as sites of competing and sometimes contradictory regulations, perceptions, opportunities, discourses and images.

The discussion that follows attempts to make sense of how these disparate trajectories may inform, and be informed by, consumer desires and practices. The discussion is based upon focus groups conducted by the authors with ordinary consumers across England, ranging from 18 to 80 years of age. The methodology for the study is described in Chapter 1. All respondents have been rendered anonymous and the names given have been chosen at random to represent gender. This chapter has, essentially, two principle objectives. The first, and larger part, is to explore the diversity of patterns and behaviours at night, while recognizing consumers as a heterogeneous and dynamic group. Second, it aims to consider the barriers that planners experience in attempting to respond to and promote diversity.

Re-framing everyday behaviour at night

This chapter will capture some of the disparity in late-night consumer experience and preference by focusing largely on everyday forms of behaviour at night. The everydayness of some late-night practices need not be seen as necessarily 'boring'. As Hubbard has argued in his *Cities*, the everyday is in fact the site for resistance and change (Hubbard 2006). At the most literal level, the simple act of walking to a local late-night store can re-inscribe the city from the ways it is often represented as a site of potential danger, chaos and binge drinking to a space that is accessible and able to be actively inscribed by the user. Such simple acts not only invert the trope of fear that has come to dominate media reports of the city, it also points to how cities are a composite of different meanings for different people, that can change throughout the day and night. Moreover, simple acts such as late-night walks, 'hanging out' or window shopping hint at the pleasurable ways urban space can be used that is not about high-risk, commercially driven or extreme forms of behaviour.

As noted in Chapter 8, planning policy has focused on retailing in town centres and there is a reticence to work with ideas of everyday pleasure. The problems witnessed in the night-time economies across Britain today can be partly attributed to this poor understanding of pleasure, risk, play (Stevens 2007), and fun, whether that be in terms of youth cultures or everyday drinking habits. This is not a criticism that has only been levelled at planning departments, but can be applied across governance more generally (O'Malley and Valverde 2004). Specific models have come to dominate the way the city at night is understood, ranging from the modern-day nostalgia for sober and sedate European-style cities, as discussed earlier, to the 'add culture and stir' model which attempts to civilize town and city centres by simply opening late-night museums, libraries and galleries. These aspirational models are of value, but they are not firmly grounded in the ways that cities are inhabited and lived, nor do they recognize either the routines of some leisure practices, or the spontaneity of others.

Even within the rubric of heavy drinking, which for some consumers forms part of their everyday existence, as argued in Chapter 5, there is a pressing need to move beyond simple explanations for consumer behaviour. Five contributory factors are typically used to explain specific forms of late-night behaviour and consumer motivations. First, the rise of youthful binge drinking has been attributed to shifting economic relations. In this account, the demise of an industrial economy centred around manufacturing to one based upon tertiary industries is believed to result in a form of social isolation, a 'crises' of masculinity and a sense of general boredom due to the dominance of low-paid and repetitive service sector employment. Other economic factors, including pricing, taxation and price-discounting promotions such as happy hours, are believed to motivate extreme drinking patterns. Second, the media is held accountable for shaping consumer behaviour, with advertising and marketing targeted at young people frequently blamed for causing binge drinking. According to this argument, consumers, particularly young consumers, are unable to negotiate media messages and, in the most reductive sense, it is argued consumers drink simply because advertisers tell them to. A third account looks at social factors such as the increased participation of women in the workforce, later age in marriage and delayed child rearing. Common to each of these factors is a greater financial freedom and independence, resulting in more time able to be spent pursuing hedonistic pleasures. Fourth, there is the political motive, which attributes legislative changes to an increase in consumption. Extended hours, a rise in the number of bars and clubs, and the increased availability of alcohol from local stores or supermarkets have allowed for much greater access. Finally, there are environmental factors, such as the design of bars, their capacity or the concentration of venues in particular hotspots and how this encourages a limited type of behaviour.

Each of these contributory factors, economic, media, social, political and environmental, plays a role in alcohol consumption and the manner in which the late-night city is experienced. Absent from these discussions, however, is an understanding of the *extent* to which each factor plays a role, in different contexts, and for different consumers. How these contributory factors coalesce differently, in different contexts, with different results and different intensities in different bodies is generally overlooked. For example, it is important to acknowledge that context and company plays a key role in our behavioural choices, and therefore, it is likely that the choice of what to do, with whom, and where will influence the actual role that advertising, for example, plays. A night out at a restaurant with one's parents may see advertising play less of a role in deciding what to drink. On an annual 'boys' night out' with a group of likeminded friends where there are different expectations around gender and what constitutes having 'fun', environmental and economic factors or the media may have a different impact. Overly tidy explanations for consumer behaviours fail to consider how these various contributory factors will operate differently, or take on a different order of importance, in different social situations, at different times of the day or night of the week.

A failure to engage with consumer practice, motivations, pleasures and desires detracts from an overall understanding of the consequences *for the night* of the urban changes that are occurring. The impact of increases in the number of higher education students living in urban centres has already been discussed. As was discussed in Chapter 2, it is no longer the case that households comprise a working father and a stay-at-home mother able to do the weekly shopping during the day. With 30 per cent of households now being single occupancy, and some central areas of London having 40 per cent single households, it is necessary for many working singletons to shop after work or on weekends. However, ordinary services continue to be geared towards outdated models of family life, and as the quotations at the beginning of this chapter attest, working singles will often have difficulties utilizing basic services such as going to the post office or the dry cleaners. The market is responding to these changes, albeit slowly.

This chapter can only begin to introduce some of the ways that people experience the city at night. We begin by addressing the fundamental issue of who today's late-night consumers actually are. In Chapter 4 it was noted that the group most likely to venture out at night are 18–24-year-olds, while the over-50s are significant in their absence. Within this broad generalization, there are several points to bear in mind. The 18–24-year-old demographic is not a homogeneous group, any more than the over-50 community is. It should also be noted that research conducted by the authors found that what actually constitutes 'going out' is not universally agreed upon. A trip to the video store after dark, or attending a gym may not be seen as officially going out at night, for example, and older people spoken to will purposefully avoid town centres but will travel long distances to country-style pubs and inns on the town or city fringe. Figures, to be explored later, also demonstrate an increase in over-50s dining out, which questions the idea that only a very small portion of this demographic venture out after dark. What is meant by 'going out' is therefore not always entirely clear and avoiding the town centre does not always mean avoiding going out to other parts of the city or even further afield.

Alcohol, leisure and fun in the modern city

At the centre of leisure activities for many people is the consumption of alcohol. This is not always the case of course, and almost 10 per cent of British people do not drink alcohol at all (Cabinet Office 2004). More importantly, the evidence presented below suggests that alcohol consumption is important, but not necessarily the only factor within a night out. Consumers do not drink in a vacuum, in other words, but for a variety of reasons and within a number of different social situations. Alcohol can be consumed as part of a relaxing dinner with a partner, or with a group of friends after a trip to the cinema. It can be consumed individually over lunch in a town centre chain bar, or in a large group and to deliberate excess as part of an annual 'girls night out'. Overall, consumers spoken to by the authors expressed disdain for excessive consumption; at least by others. Alcohol itself, however, was a key component of going out and its meanings differed according to how it was consumed, where and with whom.

As noted above, the consumption of alcohol raises a number of tensions for late-night consumers. At one and the same time, their choices are rendered as non-choices due to the powerful influence of the government or unscrupulous marketers and licensees, which render them powerless, or, they are seen as free-wheeling subjects able to take sole responsibility for their own behaviour at night. The discussion also highlighted the ways that consumers of late-night culture are often in a precarious position. Despite how the evening and night-time economies have mutated and changed over the years, the suspicions that Schlör discusses that circulated around those who ventured out in the nineteenth century continues today. Bianchini argues that to engage in late-night culture is often seen in comparable terms to a 'hobby', such as an active interest in political affairs (Bianchini 1995: 123). Unlike displaying an interest in politics, however, an interest in late-night culture does not come with the same prestige, and tends to be associated with mood-altering substances such as alcohol or drugs. There also remains a pervasive unease with late-night culture

and those who partake. Unease around youth and their corruptibility, particularly of young women, and an anxiety about urban life continues to inform the
more extreme, negative views of the night-time city.

Moving away from these moral judgements, a more practical wari-
ness about going out at night owes to the fear of assault and attack. Evidence
compiled by the authors certainly found that fear acted as a deterrent for some
consumers. However, it was not the principle reason why individuals avoided
town centres, and an element of risk was typically factored into a night out.
Consumers expressed concern for violence that may spontaneously 'go off',
but they also knew which venues to avoid, and younger people in particular felt
they were able to negotiate the city in such a way that confrontational situations could be avoided. When asked about using late-night transport, or
walking home alone in the dark, respondents said:

SARA: Yeah, I'm a bit stupid for doing it really but I do sort of just head home. I
get a homing bug and just ... when I'm drunk, yeah. If I'm a bit paranoid
and I'm listening to what's going on around me I just put my shoes in my
hand and go.
TANYA: Yeah, you take your shoes off and go. When you're drunk, though,
you're just like 'Yeah, yeah, see you later' you don't think 'Oh shit!'. D'you
know what I mean?
(Focus group member, south-western town, 2006)

Jayne *et al.* warn against painting 'an overly neat picture of evening and late-
night drinking ghettoes ... populated by drunken and violent young people'
(Jayne *et al.* 2008: 246). The late-night city is far more disparate than simply
existing under a blanket of violence and homogeneous consumers. Equally, as
Sparks *et al.* have noted, fears of crime in urban contexts can be ambivalent, as
the above quote suggests, as well as reflective more of social change than
actual risk (Sparks *et al.* 2001). For example, in one town, 'migrants' meeting
up at night outside town centre retail stores prompted fear in one middle-aged
resident. It was not clear why, however, and – following Sparks *et al.* – could
be attributed to broader social changes that had occurred in the town, increasing migration for example, than any actual negative experiences with these
newer residents.

The ambivalence consumers demonstrated about the subject of personal security at night extended to particular bars or parts of the city at different times. A concern about encountering 'attitude' deterred consumers from
attending particular venues at some points in the night, though this did not
translate into avoiding such places altogether. 'Attitude' was associated with
youthful chain bars and young, or under-age, drinkers. Again, the fear of
encountering attitude is not entirely transparent. Excessive consumption by

one's own friends or social group did not translate into an acceptance of excessive consumption by others. Equally, the notion of 'young' consumers was not entirely clear, with consumers as young as 21 years old having voiced their concerns about 'youth'.

In light of the figures cited earlier about people avoiding town centres at night and the belief that cities have become no-go areas, there have been calls to make cities more open and freely accessible. The Parliamentary Select Committee Report, *The Evening Economy and the Urban Renaissance*, for example, states that:

> Most European cities have a very inclusive evening economy where people of all ages participate in a range of activities. In contrast the evening activities of British cities are not so compatible with the inclusive ideals of the urban renaissance. They centre around young people and alcohol, leading to associated problems of crime and disorder, noise and nuisance.
>
> (House of Commons 2003b: 3)

British cities at night could indeed be far more inclusive and representative of various ages, cultures and desires. There is also nothing to justify the fear of violence associated with going out. However, it would be equally fair to note that such conditions of openness and permeability do not operate during the day. The narrative of the city as a diverse and heterogeneous place has perhaps always been a goal, but it is no more a reality of the day than it is of the night. Moreover, as Latham has argued, in order to create a diverse and heterogeneous urban environment, a critical discussion needs to be had about what this would mean, how it would work and at what social cost (Latham 2003: 1703).

Of course, the 'issue' today is that the most visible, profitable and troublesome form of nightlife is youth-centred bars and clubs. The melting pot for new forms of sociality that Sennett (2007), Somers (1971) or Latham (2003) refer to appears to be far removed from the reality of the streets of Britain, where there is a dominance of a specific type of venue aimed at a narrow segment of the community. It should not be presumed that corporately owned venues enable only one form of social interaction, or that their customers are a homogeneous group. Even in chain bars, the forms of drinking that occur vary according to different times of the day or night, different days of the week and the company that people keep. In discussions with consumers, it was also understood that chain bars were not necessarily the location for late-night disorder, but rather the spaces outside.

ALI: You don't see much in the bars. You see people getting drunk. You see the odd struggle. But it doesn't really go any further than that. What you find is when there's these big groups of people converging outside the chippies, trying to get taxis. That's when it starts, because everyone at that point, you stop drinking, you want to go back to a party or go home. So you've got all these drunken young people. Not even young people, any people. That's when it all kicks off, for some unknown reason, that you've got all these people together at the same time.

(Focus group, north-western town, 2006)

Nonetheless, though consumers attended chain bars, they rarely spoke of them in favourable terms. Instead, there was a repeated desire expressed for locally managed, diverse and multifaceted venues. Whether this will necessarily impact on the street and reduce violence is another question, but consumers spoke favourably of venues that depart from the commercially driven, youth-orientated model. We turn now to examine other forms of entertainment, alcohol-centred and otherwise, and consider the alternative forms of leisure for which consumers expressed a desire.

Community pubs

Town centre superbars that hold up to 3,000 people, and large chain venues where standing is more common than sitting, are represented in comparable ways today to the gin palaces of the Victorian period; giant venues devoted to the singular activity of getting drunk. While it can only be presumed drinking does indeed occur in these modern palaces of consumption, the variety of entertainment opportunities available in public houses in general should not be ignored. Pubs have long been used for a variety of purposes, from playing games and eating to seeking out company (Kneale 1999: 334). Ackroyd argues that there is now a wider variety of public houses than ever before, and, as well as providing the usual food and games, they cater for karaoke, dancing, quiz nights, television or live theatre. They may have gardens, private rooms and separate dining areas (Ackroyd 2001). They may serve as make-shift meeting houses for community groups as well as serving a specific section of society, from office workers to providing economical lunches and breakfasts to an older clientele.

Pubs, in both Britain and Australia, are nonetheless a subject of debate and concern. In Australia, as was explored in relation to Sydney and Melbourne, they have been accused of being rowdy and 'uncivilized'. In New South Wales the preponderance of gaming machines and televisions in the city's pubs was used to justify the introduction of legislation that allowed for small bars to open. These small bars, already found across Melbourne's central business district, are intended to enable a more comfortable social experience. In Britain, small bars are not yet on the political agenda. Instead, organizations such as CAMRA, the Campaign for Real Ale, lobby against the closure of local, community and often country pubs which are understood to be fundamentally different to the rowdy, youth-orientated, anonymous chain bars. CAMRA was initially established in response to the concentration of brewing in the hands of large brewing companies. They have since lobbied against the mass production of beer and supermarkets offering promotions of alcohol. CAMRA also support micro-breweries, while advocating minimum pricing and 'local' venues. In an argument that inverts the earlier discussion about out-of-town leisure centres competing with the city centres, local and country pubs are felt to be under threat from the concentration of venues in the city centre. CAMRA argues that up to 56 pubs close in Britain every month and, cleverly articulating pubs with British national identity, community and localism, they lobby vigorously to 'save' these pubs from the tide of globalization and cultural homogenization.

Local pubs may well be struggling in Britain, but pubs in general remain central to British leisure. As of 2007, the high-street pub market was worth almost £3 billion. This represented an increase of 1.7 per cent from 2006. Mintel's later 2008 report, *Pub Visiting*, found the most common pattern of attending a pub was weekly, with eating at a pub occurring monthly. Age was a deciding factor, with 18–24-year-olds twice as likely as any other age group to visit a pub twice per week or more (Mintel 2008). Chatterton and Hollands found a similar pattern of age dominance, with British students spending, on average, eight hours per week in pubs. It was their primary source of entertainment and more time was spent in pubs than playing sport, shopping or visiting the cinema and theatre combined (Chatterton and Hollands 2003: 130).

Determining why people visit pubs, local or otherwise, is complex. As argued earlier, rarely do people go to the pub simply to drink, and the variety of entertainment on offer in pubs is notable. In another report conducted by Mintel, *Enticement for Visiting Pubs* (2000), the reasons were, in order of popularity, entertainment (i.e. live bands, DJ, jukebox) for 33 per cent; better catering provisions for 26 per cent; sports coverage and television screens at 23 per cent; a children's area at 23 per cent; theme nights (i.e. quizzes, karaoke, BBQs) at 19 per cent; and table service at 18 per cent.

Amongst the consumers interviewed by the authors, the question was not why they attended pubs, but why not. 'Comfort' was a common refrain. It was not always entirely clear what comfort actually meant (Eldridge and Roberts 2008) and was often about a sense of camaraderie rather than sobriety. Nonetheless, comfortable pubs and bars were seen as absent in many late-night cities.

JUNE: Most of the pubs and clubs around here cater to fairly mainstream
　　　types; which is fine, I don't actually mind in the week going for a quiet pint
　　　when they have some quiet music on but I also like alternative music and

there's nowhere to go for that so I would have to go elsewhere. But, on the other hand, I do go out with my friends. We do go out around town on the weekends and I never like it. It's always lots of sweaty drunken men who fall all over you, spill beer all over you. The girlfriends then try to beat you up because you're looking at them, and oh!

(Focus group, northern town, 2005)

As discussed in Chapter 3, increasingly in Britain, pubs and bars are owned by large conglomerates where profit is the driving motive of the business (Chatterton and Hollands 2003). The demise of independent venues and the encroachment of identikit bars aimed at young drinkers has led to a sense of disquiet amongst some consumers. The sense of unease is not unique to older people, though they may typically be more vocal and feel more excluded from youthful chain bars. Young consumers also expressed a disdain for corporate-owned, city centre venues.

SANDIE: I much prefer to be able to sit down and chat when I go out. I don't want to hear blaring music in my ear and get knocked over. No matter where you stand you get pushed and shoved and get drinks splattered all over you, I'm not into that. I was twelve years ago.
SIMON: They have different names and different times, but you go into any pub on High Street and they're pretty much all the same. It's the same drinks, same attitude and they're perfectly nice places to be on a weekday or weeknight, but on the weekend they're all full of the same people just going from pub to pub because they're all the same.

(Focus group, north-west, 2005)

It is worthwhile pausing here to note that the boundaries between what is understood to be a city centre, corporate-owned, local or community pub was porous. Consumers may also be unaware of whether a venue is corporate-owned or otherwise. Research conducted in London's Shoreditch, for example, found a high proportion of corporate venues, despite the mistaken impression that the area is dominated by alternative and independent pubs (Woollett 2008). In the minds of consumers, however, local and independent pubs remain popular in Britain, with country pubs especially so. Consumers were willing to travel significant distances to attend a community-style pub and whether it was corporately owned or not was not the entire story. Common to these community pubs was a sense of inclusion and camaraderie, and the absence of loud music, drunken youth and 'attitude'. That these venues are economically viable means, however, that they are still attracting a sizable portion of the market.

In the focus groups, smoking was raised on several occasions, with parents in particular voicing a preference for smoke-free areas. This complaint is now redundant since the ban on smoking in public places came into effect in England on 1 July 2007. Though the ban on smoking in workplaces received little debate, the ban on smoking in clubs and bars was more heated. In one sense, the debate recalled the argument that the night-time city should be a place of dubious pleasures and freedom from regulation. There was also concern about elderly patrons and independent bars that were already struggling to compete with the chain venues.

Smoking has already had an impact on late-night culture in Britain, with bars and cafés occupying outside space (Bell 2008). The volley of complaints that followed the ban on smoking in public spaces in New York, which centred around noise for residents, has not yet materialized with quite the same force in the UK. Smoking outside pubs and bars in London and other metropolitan centres appears to be providing an impetus for licensees to negotiate outside seating areas and awnings, thereby lending additional animation to the street scene.

Alternatives to drinking

In exploring the various forms of non-alcohol-related entertainment desired at night, ice-skating may not appear the obvious place to begin. In the focus groups conducted across England, however, there was a repeated desire for a local ice-skating rink. This was an unexpected finding of the research. Skating can be dangerous, it is not especially economical, it tends to be cold and skating in a repetitive circle may not, in an obvious sense, appear to be especially interesting. Clearly, however, there is something about skating that warrants further investigation. Ice-skating was popular with families and thus seen as an ideal family night out. It was equally popular with younger residents living on the urban fringe, however, and across focus groups conducted in six different locations.

The venues in which ice-skating occurs are traditionally located outside the city centre, with the exception of central London where two temporary winter rinks have become firmly established outside the Natural History Museum and in the courtyard of Somerset House. More generally, rinks are located in out-of-town leisure centres, which, in Britain, are close to Hannigan's conception of 'fantasy city' (Hannigan 1998). They are marked by cinemas, amusement arcades, restaurants and bars, as well as modern gymnasiums and, common to all of them, a sea of parking. In Australia and Canada as well, these large 'sheds' in out-of-town locations are home to a modularization of stores (Hannigan 1998: 4). In Britain, one may expect to find family restaurants such as Frankie and Bennie's or the fish and chip chain Harry Ramsden's, as well as a cinema owned by a national chain, and leisure venues such as a bowling alley or gymnasium. As noted by Hubbard, and significantly for the discussion at hand, these 'multi-leisure parks' take approximately 90 per cent of their earnings after 5 p.m. (Hubbard 2005: 118).

In Maidstone, south-east Kent, Lockmeadow is anchored by a night-club that also hosts under-age events, and a range of family dining establishments. Before 9 p.m. it is dominated by families attending chain restaurants. As the night progresses a cross-section of the community come into contact as families make their way out and young people make their way in to attend the nightclub. Groups of teenagers will also congregate outside, or inside where they sit on the floor. This is an example of a space that could easily be seen as poorly planned, but equally as representative of the 'ideal' city where difference and the unexpected come into contact. Romford, to London's east, offers the Brewery, which is located just off the town's high street. Romford has acquired a reputation for late-night disorder and youthful chain bars, but the Brewery also offers consumers a large VUE multiplex, family restaurants similar to Lock-meadow and leisure opportunities such as a bowling alley and an adventure centre for under-12s. Though only a street away from Romford's bars and clubs, the Brewery attracts an older clientele, as well as families. Unlike Lock-meadow, there is less of a cross-over of uses and demographics. As a manager of a local family restaurant pointed out:

> Although we're in Romford, we're far enough away that the families can come up, they can park on the car park, they walk straight in, [and] they walk straight out again. The only time they see any of it is when they leave the car park and they might go past the lane there. And most of the time they will be leaving at times when these places aren't kicking out and that. They don't tend to see any of it – all they might do is see the rubbish or whatever is still left over in the morning but the council do tend to get most of it cleaned up ... so although we're right on the doorstep, it's far enough away and the parking and whatever, means they hardly ever come into contact, unless they want to.
>
> (Interview 2005)

For Hubbard, leisure and retail centres can challenge the urban renaissance agenda (Hubbard 2005: 119). Bluewater, in England's south-east, is one example of a suburban fringe development that is both a traditional shopping mall as well as a leisure and entertainment development. Focus groups with consumers conducted in the south-east found that Bluewater indeed draws in consumers from a wide catchment area. They are attracted to its clean environment, safety, free parking and the range of entertainment and retail opportunities on offer. Echoing Hubbard's findings on out-of-town leisure centres in Leicester (Hubbard 2005), Bluewater was also 'predictable' and a site where one was less likely to encounter the 'attitude' associated with some inner-city venues, or the chance that things could 'go off'.

The perception that leisure centres are bland, sterile and homogeneous spaces should not extend to a similarly reductive understanding of their users. As Hubbard notes, interaction with strangers may be 'essentially limited' (Hubbard 2005: 125), though Maidstone's Lockmeadow is a counter-example, but consumers are no less engaged with their surroundings, or more pliant simply because they frequent these venues. Leisure centres can come with their own set of difficulties and challenges. Respondents in a focus group in a former mining town ventured out to their local leisure centre, but poor transport connections and a poorly planned entrance that privileged car users over pedestrians meant attending the site came with added pressures. At another leisure and retail centre, parents required a high degree of skill in managing their children, shopping and prams when utilizing a poorly placed bathroom facility on the top floor of the centre. These examples point to the fact that, as noted, out-of-town leisure opportunities are not altogether 'easier' or perhaps even more convenient. Moreover, far from the disparaging view of leisure centres as soulless places dominated by chains, it was town centres that were believed to have become relatively homogenized and dominated by identikit bars. The actual practices and activities that occur in town centre bars may be diverse and enable different forms of communities and connections. Nonetheless, consumers spoken to by the authors felt that town centres at night offered limited scope for entertainment or indeed social connection. Finally, leisure centres are not necessarily used only by people who wish to avoid the town centre and its associated problems. The centres simply offer a wider array of facilities, especially for families. It is also worth stressing the point that out-of-town leisure centres did not appear to attract only one type of consumer, or that city and town centres attracted an altogether different market. People can, and do, seek out a variety of experiences that will serve a number of different purposes.

Food

Dining out has been a component of late-night culture since the eighteenth century in the UK (Olsen 1986). The expansion of contemporary restaurants and cafés in Britain, however, is a much more recent phenomena. Factors used to explain the growth of this sector include more active urban lifestyles, a later age in marriage, increased disposable incomes and a greater variety of establishments. Following the same explanations behind the growth in alcohol consumption, pricing and availability are perhaps equally relevant.

Tackling each of these in turn, the provision of dining establishments has played a large part in the urban renaissance. In similar terms to the growth of bars and pubs in town centres as a means of kick-starting the economy, eating venues have also come to be understood as playing a key role. Bell and Binnie's *What's Eating Manchester* (2005) demonstrates that new dining establishments were central to the revitalization of Manchester city centre. Elsewhere, Bell argues that 'foodatainment' and 'drinkatainment' are indeed central to 'the urban regeneration spirit' (Bell 2006: 13). While cultural regeneration has tended to favour high culture, food is demonstrated by Bell to be as important.

Dining out has played a large part in appealing to new lifestyles that are centred around city centres. As Mintel (2006b) have argued, recent decades have been witness to a growth of middle-tier dinning establishments. That is, rather than being limited to fast-food-style establishments such as McDonald's or Burger King on the one hand, or an expensive fine-dining restaurant on the other, new options are now widely available from mid-market chains such as ASK, Pizza Express and the popular Wagamamma, to the equally popular 'local' Indian, Thai and Chinese. Pub meals have also grown exponentially in recent years. Mintel's *Eating Out: Ten Year Trends* points to the growing appeal of dining outside the home. Though figures between 1996 and 2000 showed a slight drop, by 2005 the number of people choosing to forsake cooking at home for a meal out had increased significantly (Mintel 2006b). Overall, since 1995, the eating-out market in Britain has increased two-thirds, to an annual value of £27.6 billion. This equates to a real-term growth of 29 per cent.

As noted, dining out, as a key component of the late-night city, has been driven by a series of interconnected economic, social and cultural changes. As well as 'benign economic conditions', Mintel's research points to the increase in women working outside the home, resulting in greater time pressures regarding cooking, a growth of inexpensive dining options and the perception that dining out is now a common and normal part of urban and sub-urban lifestyles (Mintel 2006a). Of the respondents interviewed, 96.1 per cent surveyed by Mintel ate out, which, 'to put this in perspective ... means that eating out is now as common as watching television'. Two particularly striking points in the Mintel survey are, first, that the rate in which the social group ABs (professional and managerial class) eat out has slowed. Though they still make up the major proportion of people who dine out, C1s have grown faster than the general population; a trend attributed, as discussed above, to the increase in mid-market dining establishments. The second noteworthy point is the pre-ponderance of 35–44-year-olds who regularly eat out, and are classified as 'heavy users'. This group represents 21.6 per cent of 'heavy users'; and though family responsibilities lead to a marked difference amongst this group, with parents eating out less than non-parents, Mintel note the 'great strides' taken by the hospitality sector in the growth of family-friendly dining establishments between 1996 and 2005. A further noted trend is the increase in 'light-users', a trend attributed to older women, with the over-55 market being the highest percentage of all 'light users'.

These 'great strides' in attracting families to eat out at night were a common refrain amongst supporters of England's 2007 ban on smoking in restaurants and bars. The ban, which affected all public spaces including bars, restaurants and nightclubs, was frequently justified in terms of the growth of dining out. The Mintel study cited above notes that the pub-dining trend grew substantially between 1996 and 2006, with the gastropub (venues which offer fine dining in an upmarket pub environment) leading the way. Of meals taken outside the home, 23 per cent now take place in pubs (Mintel Summary 2006).

Thinking specifically of the restaurant rather than pub sector, in 2005 the market was worth slightly under £6 billion. Of this total, one-third was attributed to dining in ethnic restaurants, with Indian and Chinese remaining the most popular (Mintel 2006b). The popularity of Indian, Chinese and Thai food in Britain highlights the extent to which Britain's evening economy and consumer practices have changed from traditional local 'boozers' and greasy spoons. Socially diverse establishments that attract all types of consumers, ranging across age, gender, race and class, may not be entirely common, but then nightlife, and activities during the day-time, are also marked by difference in terms of these axes.

Class and age, as demonstrated by the Mintel figures, are important variables in the uptake of the 'food renaissance'. However, a finding of our research was a striking commonality about the centrality of dining out to the experience of going out. At the high end of the market, there are organic farmers' markets and gastropubs, with working-class respondents more favourable towards family-style establishments. Overall, however, the research found that 'decent food' was a common refrain amongst all respondents. In these terms, it was a better class of establishment that was often called for. In one study area, a former mining town, this manifested in the desire for little more than an up-market high-street pizzeria. In a town in the south-east, in contrast, consumers expressed a desire for restaurants with 'table linen'. The two types of establishments suggested are vastly different, but they both represent a comparable wish for a more aspirational dining experience.

Childcare and families

A report commissioned for the Joseph Rowntree Foundation (Statham and Mooney 2002) examines the growth of late-night childcare. The researchers from the Thomas Coram Research Unit who conducted the study found that while there is a recognized need for late-night childcare, only one-third of providers surveyed in the UK had attempted to offer such a service. Instead, they found a demand for childcare in the early evening, rather than late or overnight, and much of this was being met by family, partners or friends. A further significant finding of the study was that new forms of childcare offered in atypical hours, that is, outside of 8 a.m. to 6 p.m. Monday to Friday, was only recently beginning to emerge.

For the individuals spoken to in our research, childcare, or the lack thereof, was an impediment to venturing out more often. However, this was not a 'problem' that could be easily solved. There was a reluctance to employ child minders, or to place children in childcare establishments. Concerns over the child's safety were common, but it was also noted that with both parents working, there were few opportunities for families to be together. And, with many restaurants being unsuitable, home was the obvious location for socializing and bonding. As one parent noted:

DAMON: I get out fairly rarely. I've got a young child and it's just a military operation trying to organize baby sitters. It's a fairly rare occurrence. It has to be when relatives are visiting and then you get your annual night out.

(Focus group, south-western town, 2005)

As well as the effort required to organize child care, or to prepare children for a night out, there is a perceived lack of suitable spaces for 'families' in the UK at night. This does not mean there are no child-friendly venues; places such as various well-known chain restaurants were recognized as suitable and safe venues for children. But these venues were only 'child-friendly', not also adult-friendly, meaning adults were there merely as chaperones. Venues where adults could have a drink and good food while children could play were largely lacking. A further and often un-remarked upon problem was that the parents spoken to did not have children of similar ages. It was far more likely to have children ranging from infants to teenagers. Finding venues that appealed to the entire family, including adults, was challenging. One alternative frequently raised was festivals and markets. As we have explored elsewhere, the desire for festivals and markets may appear antithetical to the corresponding dislike for crowded, noisy venues (Eldridge and Roberts 2008). Common to both experiences, however, is a sense of camaraderie and belonging.

One example where the introduction of a festival widened the demographic coming to the town centre at night comes from Newcastle-under-Lyme. This historic market town had been overtaken by an expansion of youth drinking venues. The Borough Council was persuaded to contribute £10,000 for a jazz and blues festival in 2006, an event that was supported by local licensees. The event proved to be popular, drawing an older, more middle-aged clientele into the town centre during the day and in the evening. Subsequently, a small number of the traditional pubs were inspired to hold jazz nights as a regular weekly event. They were able, as a result, to start their evenings earlier and close earlier because of the older clientele. The three-day festival has now become an annual event. Its success prompted the council's former chief executive to comment that this 'soft' measure to change the ambience of the town was as effective as 'harder' measures such as policing.

Shopping

A final form of leisure discussed with consumers was shopping. Shopping in Britain today is perhaps the clearest example of where day-to-day activities have extended into the night (Blau 2008). Tesco, the giant international retail chain, opened its first 24-hour store in 1998. Of its 400 'Extra' stores, which are typically out-of-town hypermarkets, 300 are now open 24 hours per day from Monday to Saturday, and 10 a.m. to 4 p.m. on Sundays (Cottam 2006: 26). Sainsbury's Express or Tesco Metro, small urban stores, are also now open late, typically closing at midnight. According to Cottam, '67% of consumers see around-the-clock supermarket opening as a positive trend, and about 15% of shopping now takes place between the hours of 6 p.m.–9 p.m.'. Geiger notes that 17 million people now shop at night, while 58 per cent of young people without dependants expressed a desire to do so, if facilities were available (Geiger 2007: 25).

The reasons for shopping later into the night can be attributed to a series of social and economic changes concerning employment, the home life and division of domestic chores, the growth of single households and busy life-styles. On the one hand, there is everyday shopping which entails purchasing everyday items from the supermarket. Accessing services such as banks and post-offices, to carry out mundane tasks, has also been noted. On the other hand, there is a more aspirational form of shopping. London's Covent Garden was held up as one example, with its street entertainment, dining, retail and more common leisure opportunities such as pubs and bars, it was perceived to be the ideal late-night space.

London enjoys unique advantages over other locations in the UK and its size and density mean that late-night shopping can attract sufficient customers. Other smaller centres experience more difficulties in ensuring sufficient footfall to justify later opening. Nevertheless, where some degree of late-night or evening shopping had been achieved, as in our 'office town' in the south of England, there was evidence that it was well received. In this town centre, the developer of the shopping centre concerned, in consultation with the council, had agreed that a condition of the retail leases would be that the individual units stayed open until 8 p.m. in the evening. As well as retail opportunities, the centre also offered exercise facilities, restaurants, bars and a cinema.

AUDREY: It's made a big difference.

BARRY: It has, it has made the town centre a much pleasanter place to be in the earlier evening.

JENNY: They go swimming early in the evening and then they might say, 'Let me pick the children up', then they go onto the theatre or see a film. It's quite good really.

GRAHAM: I think on the whole that has worked.

RUTH: Now with the longer shopping period that then overlaps into the night time I don't have any issue about walking around the town centre on my own prior to going to the theatre or concert.

(Focus group, office town, 2006)

This type of initiative is relatively unusual in a provincial British context. Secondary centres such as the mining town are still struggling to fill the gap between 5.30 p.m. and 8.30 p.m.:

SHEILA: From 5.30 the town empties and the only thing it's short of is the tumbleweed blowing through it because the shoppers and office people have gone home and the town is like a dead-town and the market area is incredibly horrible and makes you feel very lonely. If you walk from the station around 7.00 or 8.00 at night, in the winter time particularly, the wind howls down the street.

(Focus group, northern town, 2005)

Planning policy and consumer choice

Arguments about deficiencies in the overall 'offer' in the evening and night-time economy mirror much of the debate that is taking place in contemporary Britain about retailing. In the absence of easily accessible research detailing the interaction of the property market and entertainment uses, it is useful to turn to this debate. One common point of concern is the loss of small independent traders, which in retailing refers to the local butcher, greengrocer or hardware shop, and in entertainment, the local pub. In each of these sectors the loss has been estimated as at least one outlet per week closing down (House of Commons 2006), with local pubs recently accelerating towards one per day (BBPA 2008). A corollary of the loss of small independents is a perception of narrowing choice in the high street (Conisbee *et al.* 2004). The domination of the major supermarkets and of multiple retailers is an issue for grocery and other shopping. This chapter and Chapters 3 and 4 have outlined concerns about the night-time high street being dominated by chain bars, pubs and clubs (Hollands and Chatterton 2003).

The common thread between the two sectors is national planning policy. In England and Wales this is shaped by *Planning Policy Statement 6: Planning for Town Centres* (PPS6) (DCLG 2005b), which, together with more general national planning guidance on principles (DCLG 2005a), sets out a 'town centres first' approach, whereby major developments in retailing and leisure are directed towards town centres. A document with less legal status within the British planning system, but which nevertheless holds some sway, guides tourism development (DCLG 2006b) – including pubs, bars, nightclubs and restaurants – towards town centres. These policies were initially developed in the mid-1990s, in an effort to direct development away from out-of-town schemes such as retail parks and leisure centres, and to revitalize town and city centres. A recent consultation document (DCLG 2008) on PPS6 notes that the technical requirements engendered by these policy documents have led to some unforeseen effects with regard to retailing. These are a limitation of consumer choice and competition, particularly in edge-of-town locations. The same report also criticizes some authorities for drawing the boundaries of their town centres too narrowly.

It seems feasible that the disappearance of independent local pubs from the high street and their replacement by chain and theme bars with a high density of sales per square metre of floor space is driven not only by changes in the structure of the operating industry, but also by planning policies. As high street rents move ever higher, local independents are driven out and there are fewer available sites on which they could re-locate. In this manner, a perverse

9.1
Closed clubs appear hostile in the day time.
Source: photograph by authors.

by-product of 'town centre first' policies is to reduce the individual identity of historic centres as local taverns and corner pubs close or become 'themed'.

The analogy between the retail and entertainment sectors only stretches so far. Retailing occupies a large and necessary part of consumer expenditure in comparison with entertainment. Different players provide investment and development, and the structure of leaseholds and tenancies differ. There are, however, points of congruence in terms of larger mixed-use schemes where a major investor develops a site that incorporates retailing, leisure and entertainment. Westfield's major new shopping centre in London's White City is an example that opened in 2008. It incorporates 47 different dining outlets alongside its 150,000 m² shopping area, with a 14-screen cinema complex projected to open in 2009. This type of large-scale development is, of course, unusual in its scale and is unlikely to be repeated.

In the pub sector the recent closures of local and neighbourhood pubs in urban areas are not easily explained. The British Beer and Pub Association ascribes their members' difficulties towards 'spiralling costs, sinking sales, fragile consumer confidence and the impact of the smoking ban' (BBPA 2008: 1). Certainly, as discussed in Chapter 5, pricing must have an impact in that it is much cheaper to drink at home before going out and, indeed, for some consumers pre-loading is an important part of the ritual of 'going out'.

Interviews by the authors with corporate providers found some of them expected the neighbourhood pub to undergo a renaissance following the implementation of the Licensing Act 2003. When interviewed prior to the Act's implementation, corporate operators felt the Act would provide an opportunity to set up smaller, quieter venues for an older, more discerning clientele. These 'chill-out' or wind-down bars might also cater for people coming out of clubs.

> The way that the market is moving is, in my eyes, it wants to move more into a café culture, sort of, aspirational, quite affluent, trendy premium offers market.... I think there is going to be less of the flagship battleship boozers, the massive banks turned into big pubs etc [and] there will be more café/bar/bistro-type operators rather than massive boozers.
>
> (Senior manager, pub chain)

> I think you are going to get smaller venues being more attractive, because one of the things about the late-night licences I think you [need] to offer people the opportunity to go out and have a drink and a chat with friends late. The problem at the moment [2005] is if you want to open late you've got to have music and food and all the rest of it, so inevitably most places are noisy and they've got dance music or music on and lights and all the rest of it, because they have

> to do that to operate their late licence. I think that you're going to find more of the old style nightclubs where you've got background music and people sit down and have a chat and talk to each other.
>
> (Senior executive, pub and club operator)

The possibility for more relaxed venues also existed for attracting older people into nightclubs. At a conference organized by the Civic Trust, one speaker suggested that a new market was waiting ready to be tapped, of a potential clientele who were returning to the clubbing scene after having established a household. The group of 40–50-year-olds he dubbed the 'virties' or virtual-30s (Collins 2004).

The fact that this has not happened suggests a topic for further research. More needs to be known about the pressures on local, independent venues in British urban centres if the aspiration set out in PPS6 for diversity in the evening and night-time economy is to be achieved. This research would need to move beyond the location of the town or city centre to the neighbourhood in order to fully understand the interactions of the property market, planning policies and consumer behaviour.

Concluding comments

This chapter has suggested that continuing concerns that surround late-night culture and the tendency towards an ahistorical understanding of drinking practices have obscured understandings of the dynamics of the evening and night-time economies. The mismatch between changing consumer practices and preferences and the types of establishments that are profiting from increased demand appears to be significant. With the *caveat* that more research is required, there is much scope for re-imagining a different, more rounded evening and night-time economy. In order to achieve this, planning and management will need to be guided towards a more diverse and alternative set of frameworks that move beyond the discourses of binge drinking, public health, and 'efficiency and competition'. In the next chapter we shall re-scrutinize our contemporary models for the evening and the late-night economy and suggest avenues for change.

Chapter 10

Night-time cities, night-time futures

> If I was urging anything, I always urge our ministers, not just in this depart-
> ment, if I am in a policy discussion anywhere, 'You can't just ignore the com-
> plexities, there are no simple things, it is complex and it takes a long time to
> find the solution.'
>
> Senior civil servant, DCMS

The first part of this chapter entails a review of key late-night narratives, how
they have surfaced and shaped production, regulation and consumption. The
current economic downturn looks set to provide a different context for the
development of night-time activities and we shall explore different scenarios.
The discussion then turns to considering some ideas for practical interventions
to improve the regulatory environment for alcohol-related entertainment.
Finally, we shall review the gaps in knowledge and suggest some avenues for
future research.

Competing narratives

The idea that the night-time is framed, developed and 'performed' in a different
manner to the quotidian activities of the day is a recurring theme. The night has
long been associated with pleasure, transgression and freedom. This can take
many forms, from the 'comfort' sought by friends who want an intimate con-
versation in a quiet corner, through to the 'risky pleasures' associated with rave
and club culture (Hutton 2006). The expectation of pleasure is a counterpoint
and acts directly in opposition to day-time activities; activities traditionally

understood as relating to the everyday worlds of labour, business and finance. Yet, simultaneous with the development of the night-time economy in the UK, there has been another trend. Everyday activities, such as convenience shopping or attending a gym, have gradually expanded further into the evening. Eating out has become more commonplace as disposable incomes have risen and household structures have changed. Working hours are changing and working into the evening is a feature of some occupations. Discussion of the night-time has needed to accommodate these two trends. Though they do not necessarily act in synergy, they are not necessarily in opposition.

Alcohol and the manner of its consumption underlies much of our discussion of the changing nature of night-time activities. That alcohol is an addictive drug renders its consumption as dangerously different from, say, fast-food. The violent crime and disorder that are associated with numbers of inebriated people coming out onto the streets simultaneously cannot, as we have argued, be ignored. The idea of a 'moral panic' is difficult to examine since, on the one hand, there is a wealth of statistics that reveal private tragedies such as assaults and even murder. To subsequently dismiss concerns about alcohol-related disorder as merely a 'moral panic' can also obscure very real problems. The notion of 'liveable cities' is undermined by micro-districts becoming 'no-go' areas late at night. On the other hand, the moral panic about 'Binge Britain' can obscure other no-less-serious alcohol-related problems associated with, for example, middle-class consumption, violence within the home or, indeed, the ways that some people may play up to and appropriate the moral panic message.

At the same time the notion that society is made up of individuals who freely choose how and what to consume, unfettered by any considerations other than immediate desire, is equally a misapprehension. 'Boosterism', 24-hour cities, the 'cultural turn', the relaxation of licensing hours, changes in youth culture, the expansion of higher education and changes in the drinks industry have contributed towards a steady expansion in late-night, alcohol-related entertainment – 'drinkatainment'. This was driven by competition and shaped by the property market towards the most profitable establishments, large-scale vertical drinking that would attract a youthful clientele. The relationship between supply and demand is complicated. In short, the framing of a 'good night out' is not outside an individual's control, but is also not determined by it.

Tensions over the precise nature of responsibility and individual choice are played out over a number of arenas. Within government there are tensions between different ministries, with the DCMS wishing to boost the hospitality industries, the Home Office to restrain behaviour and the DCLG to mediate between the two. Central government relies on local government to carry out its policies, and local authorities have to take a pragmatic course veering between laissez-faire and strong interventionist approaches. The criminal justice system similarly veers between the permissive and the repressive, on the one hand not enforcing a law that it is illegal to serve somebody who is already drunk, and, on the other, defining more and more statutes that control individual consumer behaviour. Within the private sector the 'orderly' operators blame the poorly managed, while an independent study points to widespread disregard for standards of good practice. Meanwhile, clubs, pub and bar operators blame cheap supermarket deals for violence and disorder while the major supermarkets insist that they serve a day-time market and night-time disorder is beyond their remit. The rhetoric of responsibility, individualism and the public and private forms an important backdrop to discussion of alcohol, regulation and consumption. The limits to liberalism, whether in terms of neo-liberal economics or 'liberal restraint' are tested by the expansion of the night-time economy. A further trope is that of a 'continental' Other, the idea of an alternative style of drinking that is civilized and restrained, yet somehow unregulated. This, too, has been demonstrated to be a myth.

Our argument is that an oppositional and aspirational discourse about the night-time economy serves to mask its uneven development and its relation to the everyday. The determined pursuit of pleasure, in which alcohol is a component, distinguishes the territory of the night from that of the day. This is not to deny that with pleasure comes danger and fear. Nevertheless, night-time activities are still subject to the same determinants as are day-time, in that profits have to be made, customers found, people transported and litter cleared away. To complicate matters further, day-time activities are now interjecting more firmly into the night. The task for planners and other built-environment professionals, is to respond to both the night's extraordinary properties and its everyday requirements in an appropriate measure.

Entrepreneurialism, managerialism and collaborative approaches

The expansion of the alcohol-related night-time economy has been characterized as indicative of the rise of an 'entrepreneurial' approach to governance characterized by the adoption of public–private partnerships. As the preceding chapters have demonstrated, this is a somewhat misleading interpretation of events, at least in Britain. The leisure-focused public–private partnerships of the 1980s and 1990s produced out-of-town sanitized environments in which people went to the cinema, not outrageous clubs in derelict warehouses. The narrative that characterized the growth of nightlife in the 1990s was that of straightforward capitalist accumulation, with corporate and independent operators taking opportunities wherever they could find them and using their financial resources to quash local opposition. The local state only became involved as a partner when the negative externalities associated with excessive competition and concentration threatened public order. At this point, different types of pub-lic–private partnerships were established, in the form of licensing forums, night-time economy working groups and so on. The brief of these organizations is 'boosterist', with Nottingham's night-time economy BID providing the most explicit example of the 'entrepreneurial city' ethos.

Central government has resisted direct intervention into suppress-ing the supply of licensed venues. Instead, it has pursued a course of vilifying errant consumers and 'rogue' licensees, arguing that corporate and social responsibility will resolve questions of over-consumption and public (dis)order. When faced with a sustained campaign against 'Binge Britain', public adminis-tration drew on the ethos of collaborative planning and public participation by allowing some leeway by local community groups although, paradoxically, pre-venting ward councillors as locally elected representatives from making objec-tions. While this encouragement for local control is admirable in many respects, it relies on the availability of an informed and articulate local citizenry and busi-ness community.

This is not to deny that an entrepreneurial approach of partnership working at the local level cannot be commended. The Civic Trust's purple flag scheme attempts a higher level of development, setting out a ten-point stand-ard for managing the night-time 'offer'. It includes responding to customer need, the encouragement of mixed-use and nurturing new types of culture and venues. The scheme bears a resemblance to Jim Peter's work in the USA, which provides a similar consultancy service for towns that have ailing or prob-lematic night-time economies. Both schemes are research-based, sensitive to context and broad-minded. Undoubtedly, the flag scheme offers sound advice that will help to reduce crime and attract new people into urban centres. Yet there remains the sneaking suspicion that the essential point about a 'vibrant culture' is that it resists such packaging. Fiona Measham's study of a club down a dark stairwell that plays authentic drum and bass music on a Thursday night is a quintessential example of the creativity of nightlife. However, that same club is highly unlikely to be signing up to pubwatch or sending represent-atives to the local forum.

The entrepreneurial approach is also subject to criticisms of gentrifi-cation. As an illustration, the newly revitalized South Bank Centre in London was the first recipient of a 'Nightvision' award. Undoubtedly, the refurbishment of the undercroft of the Royal Festival Hall, with chain restaurants and shops and the construction of other similar establishments nearby has proved to be a popular success. The associated riverside walk provides a magical, openly accessible public space that requires no entrance fee or dress code to enjoy. Even though architectural critics sneer at the 'tacky' shops (Glancey 2008), the food and drink 'offer' still lies outside the reach of low-income groups, many of whom have secure tenancies on council estates within walking distance of the centre.

While neo-liberal, entrepreneurial and collaborative approaches each have their weaknesses, it is clear that nightlife defies the 'command and control' economy. The 'greyness' of life in the former communist states was a consistent source of criticism, and while the beer may have been cheap, the circumstances of its serving did not contribute to a sense of universal *joie de vivre*. As the saying went, it was preferable to work under communism and live under capitalism. Entertainment is paradoxical in that its variation and uneven development requires elements of entrepreneurial, managerialist, collaborative and neo-liberal approaches to produce the outcomes that deliver the most fruitful outcomes at a local level, tempered by social democratic regulation. Neo-liberalism created the circumstances where new forms of youth entertainment and culture could flourish in the early 1990s. The managerialist approach that Westminster City Council employed in response to over-concentration led the way in defining cumulative impact zones and controlling the powerful interests at work in London's West End. Conversely, entrepreneurial approaches have assisted in the development of declining towns such as Doncaster as regional centres for entertainment. A collaborative approach involving a powerful residents' grouping has helped to resolve differences in Norwich.

At a national level, the central state wields more power and can intervene effectively. So far, it seems unwilling to curb the dynamic of growth, except in specified local circumstances. To a certain extent, central government's hands are tied. European law prevents imposing higher taxes, and therefore the price of alcohol being raised by any significant level is unlikely. Nevertheless, changes to competition law and to national legislation and policy could change the rules of the game under which the nightlife industry operates. This book has supported the argument, advanced by others, that the unrestricted growth of alcohol-related entertainment reveals the inadequacy of market mechanisms to promote the public good. There is a strong case for central and local government to exert more control over the distribution, size and style of venues. Alcohol disorder zones merely claw back some of the costs of excessive competition between different parts of the industry and do not contribute to wider social and environmental goals.

As was noted in Chapter 8, planners in Britain have tended to ignore entertainment as a viable land-use. Other aspects of the government's own agenda challenges this stance. Section 17 of the Crime and Disorder Act 1998 places an obligation on local authorities to reduce crime in their area and this applies to spatial planning. The re-adoption of spatial planning has the implication of allowing other issues, such as public health, to enter into the equation. If the reduction of obesity can be a legitimate objective of local plans, then the prevention of the hazardous consumption of alcohol must surely be. The debate about personal versus social responsibility has an ethical dimension. Regeneration based on excessive alcohol consumption by young people cannot accord with social justice. Even on economic grounds it is dubious, once the costs to the National Health Service and the police are taken into account.

The experience of licensing liberalization in Britain has not led to more inclusive town centres. It is easy to foresee a situation where town centres at night become a patchwork of 'no-go' areas dominated by vertical drinking establishments and deserted shopping streets with the possibility of one very small gentrified cultural zone centred around up-market restaurants, a theatre or cinema. The prices of town centre rents drive all but the most profitable venues out of the centre and those at the margins use a variety of means, such as lap dancing, to attract custom. Meanwhile, the hike in real estate has led to the closure of neighbourhood pubs and café bars, so that people are forced to drink at home. This may be a worst case scenario, but elements of this picture are present in many towns and cities across Britain. As authors, we are not experts in planning theory or business management. Our research has, however, led us towards some pragmatic observations that may be helpful in reforming the processes and practices of planning and licensing.

Everyday convenience and informal practices

In Chapters 2 and 9 we argued that the night-time city is developing unevenly, with extensions of day-time activities into the evening in some metropolitan centres. We also suggested that there is a demand for a wider variety of styles of consumption and activity than are currently being supported, either through practice or market mechanisms. Supporting this desire challenges some current practices, policies and ideas.

For example, it is conventional to regard run-down industrial areas as a series of development sites ripe for regeneration. Yet, these same industrial buildings have frequently supplied important and appropriate spaces for clubs to start up. Jane Jacobs extolled the virtues of providing a mix of age and price of building in any neighbourhood as a means to create vitality. This seems to be one of her observations that has been overlooked in assuming a 'win–win' synergy between social inclusion and economic growth. The over-determination of spatial frameworks in masterplanning practice in the UK seems likely to drive out creative activity rather than nurture it. Cultural geographers have long made a plea for the importance of 'marginal spaces' or 'incomplete form' (Sennett 2007). Regrettably, this has not been incorporated into contemporary urban design practice, which still over-emphasizes the aesthetic and financially profitable.

10.1
A market in Barcelona at 9 p.m. on a Sunday night. This demonstrates the extent to which the everyday can be inserted into night-time activities.
Source: photograph by authors.

Attitudes towards pedestrianization and the car also need re-visiting in the context of evening and night-time activities. The preceding chapters have demonstrated that transport is important. Outside of the major conurbations it is unrealistic to expect bus operators to put on a commercially viable service late in the evening and at night. Bus services also do not resolve all transport problems for women given that they are not door-to-door. Taxis and mini-cabs rely on a supply of drivers and this cannot be guaranteed. Policies of providing car-free zones in town centres and of shutting park-and-ride schemes early in the evening therefore seem particularly misguided. Families are unlikely to make an effort to come into urban centres if they have to walk a long way from their car, are uncertain about its safety and have to pay high parking charges. While a reduction in car use is a laudable objective for urban design and planning, its blanket application, in the absence of alternatives, threatens other social aspects of sustainability.

A further dimension to transport is the regional catchment of major nightclubs. In the course of our research we have come across nightclubs in Newmarket whose customers came from Ipswich, in Shoreditch in London where coaches arrived from Essex, and in Bournemouth and Brighton whose customer base extended back to London and parts of Kent. The local vision of the urban renaissance, with its notion of neighbourhood-based cafés and bars, is at odds with the reality. Integrating transport for night-time uses requires a regional overview, combined with local provision for coach parking, unloading and boarding.

The extent to which venues are concentrated or dispersed is key to a successful night-time economy. Chapters 4, 6, 7 and 8 have illustrated the problems associated with an over-concentration of venues in terms of gathering large numbers of drunken people together at one time. 'Drinking streets' discourage older people from using them and other types of night-time uses, such as smart restaurants, from setting up. The practice of 'circuiting' between venues has also been discussed. Movement between different types of venues, combined with the sounds coming from the inside, help to create a party atmosphere. From this point of view, a certain level of focusing within a group of streets is helpful. Furthermore, grouping also assists in establishing BID-type organizations which means that the costs of extra management falls to the businesses rather than the council-tax payer. The type of concentration that is to be avoided is where two large, say 2,000 or more customers, clubs either adjoin or face each other, surrounded by late-night fast-food takeaways.

It is interesting that Britain's major nightclub chain, Luminar, is now developing venues where different styles of dancing, eating and drinking all come under the same roof. If this is a significant trend, then large 'superclubs' will pose the same problems to urban designers and planners as their day-time counterparts, the shopping mall. Whereas day-time shopping malls form impenetrable barriers to movement after the shops close and present the pedestrian with blank walls and a sea of car parking, a similar phenomenon will occur around superclubs, only during the day. Moving customers to an internal space moves security issues into the private domain, which may be an argument that pleases local crime and disorder partnerships. Conversely, the removal of large numbers of people from the streets (for example, the capacity of the Basildon Liquid & Envy is 3,470 (Anonymous 2008)) re-invigorates debates about revitalizing 'dead' town centres.

At the time of writing the UK system of planning and licensing has limited scope to effect either concentration or dispersal of licensed premises. Its powers are mainly negative and proliferation can be prevented through both the planning and licensing regimes, as has been discussed in Chapter 8. The experience of positive planning, as with the Temple Bar neighbourhood in Dublin, the Riverside district in Norwich or Gracia in Barcelona, seems to have led to an over-concentration, until the policy was reversed. A further problem lies with calls for a variety of different types of eating and drinking experience. The new spatial planning system in England and Wales offers a means of directing investment, but the level of detail required by the supplementary planning documents and the time taken to prepare them means that this is an untested route. The volatility of nightlife and the way in which fashions change means that the social democratic processes of town planning, with their consultation periods and adjustments, often lag behind development.

The UK system of separating planning from licensing, as Chapter 6 discussed, does not help local authorities with forward planning. A greater integration between the two systems would assist better feedback between the direction of development on the ground and plan preparation. There is a further link between the two systems that has received little attention to date. This is the relationship between the asset value of entertainment premises and other types of land-use. One of the major pubcos in the UK had a deal with an equity fund based on the asset value of their property narrowly fall-through in 2007. A better understanding of the real estate value of licensed entertainment premises in relation to other land-uses would give greater insight into the dynamics of the industry.

Although officials in central government might argue that it is the duty of local government to integrate licensing and planning, clearer guidance at national level would undoubtedly be helpful as to procedures. The national guidance document for planning town centres, PPS6, deals briefly with the night-time economy. A more detailed treatment again would be helpful, perhaps with examples of how to achieve a diversity of offerings through a combination of different policies.

New types of planning rule might be suggested. Now that magistrates no longer define the 'need' for the number of licensed premises in any one area and licensing authorities can only refuse an application if one of the four licensing objectives will be breached, it falls to the planning system to intervene more firmly in establishing the character of different types of urban centre. The experience of European neighbours, as Chapter 7 demonstrated, is various. There are different styles of definition available. Berlin and Copenhagen use a percentage of ground-floor uses, combined with other controls. France has a measure based on customer spaces per head of the local population. This may be difficult to apply to a large superclub, where the catchment area lies across other centres. There is also the idea of placing minimum distances between, say, licensed premises and other sensitive land-uses, such as schools. Percentage limits might be imposed on the different entertainment land-use classes within individual streets.

National licensing legislation could be reformed too. Scotland has included public health within its licensing objectives and has Licensing Boards and distinct Licensing Forums that now have the powers to assess 'need'. Clearly further research would be helpful in exploring these different systems of control and establishing their wider applicability. However, an assessment of an area covered by a Licensing Forum would seem to allow broad-brush controls without excluding either inoffensive or imaginative new developments.

The focus of interest in this book and in policy is the urban centre. Understanding the dynamic of nightlife needs to be extended further outwards, into the local neighbourhood and housing estate. Competition in the industry, as has been noted, has led to the demise of local pubs. There are many reasons to regret this, despite their history as male-dominated sites for encounter. The run-down or cheap neighbourhood bar can serve many functions, as a meeting place, a respite from home, or to use Oldenberg's term, 'a third place' (Oldenburg 1999). It can also function as a place where young people learn about alcohol, comportment and consumption through the informal control mechanisms provided by family, friends and neighbours. It is commonplace in plans for sustainable development to include a site for a pub or café bar in neighbourhoods with a population of 5,000. The mechanism of pricing and land values in Britain seems unable to support this. Central government might do better to consider how to work with the industry to revitalize women-friendly and family-friendly neighbourhood pubs, bars and clubs rather than pouring resources into policing and enforcement. Encouraging people to consume outside the home has the potential to create a sense of community and to re-establish behavioural norms.

Another way in which the sense of community can be supported is through public events and festivals, as discussed in Chapters 4 and 9. The arguments made by Bianchini, Montgomery and Worpole are still highly relevant to urban areas. Regrettably, festivals and events have been incorporated into the entrepreneurial narrative, and with sponsorship by major media and drinks companies they are increasingly dominated by commercial imperatives; by public and private sectors alike. Alternatively, Leeds City Council has shown an imaginative approach to the scheduling of night-time events. As a major city, however, it has resources beyond those of many areas. Here a more positive approach could be taken by central government through the use of National Lottery funds. These speculative suggestions are taking the discussion into the future, where it cannot be assumed that current trends will automatically continue.

The future

The recent expansion of nightlife in Great Britain, Europe, Australia and parts of the USA has occurred during a period of economic expansion. At the time of writing, economies across the globe are in a downturn, if not full-blown recession. In this scenario it seems probable that town centres will be quieter at night, with less tourists, business visitors and overall activity. Some operations may be forced into closure, for example high-street restaurants and nightclubs, establishments where the overheads are high and a reduction in throughput threatens their viability. Although there may be a reduction and contraction in nightlife, it seems unlikely that it will revert to the situation of the early 1980s: the underlying pressures and changes discussed in Chapter 2 remain the same. The population of city centres has also increased with the recent spate of flat building. While the structure of ownership may change from owned to rented, there will be an opportunity for higher education institutions, registered social landlords, local authorities and other institutions to take over properties from developers, as happened in the downturn of the early 1990s.

A downturn in nightlife activity could prove to be beneficial to British urban centres. It provides the opportunity for a rebalancing and redistribution of land-uses throughout urban centres because, with an increased resident population, the demand for food shopping and other everyday conveniences will still be buoyant. If a reduction in commercial rents and asset values occurs, new opportunities will be available for cultural entrepreneurs. New spaces may become vacant in the suburbs and local neighbourhoods as housing sites are abandoned. It seems probable that the major corporate interests will continue to turn their attention to the newly emerging economies in the Pacific Rim, China and Russia. This, too, may provide opportunities in the UK for local entrepreneurs to start new ventures.

Research

The discussion above has indicated that there are several areas where more research is required. A deeper understanding of how the dynamics of the property market interact with the development of different categories of nightlife establishments is needed in order to understand underlying trends. Wider consumer surveys are required to gain a better understanding of preferences for entertainment, culture and the possibilities of influencing these. In Chapter 2 we indicated that an increase in women going out at night constituted a major change in nightlife. To date there has been little research of this phenomenon. Although the much-heralded shift towards more flexible hours and different styles of working have not been in evidence, nevertheless working patterns and lives have changed. The growth of small supermarkets and convenience stores in city centres are suggestive of different styles of living and working. The major museums and art galleries in London have started to open later on selected evenings. Gyms, leisure faculties and certain services, such as walk-in health centres, have started to experiment with opening later into the evening. More needs to be known about their customer base and how lifestyles are changing for different groups.

This book has not considered prostitution and drug use, except in passing. The dramatic recent growth in lap dancing and 'exotic' dancing reveals the depth of this association. The literature on drug use has also tended to stand apart from studies into alcohol and violence. Here again, a more nuanced understanding might help to provide insights into the threats or challenges posed in certain street environments.

Although this book has highlighted certain aspects of overseas regulation and practices, a wider international perspective on nightlife and night-time activities would help regulators and academics. The UN, the WHO and EuroCare have each carried out important studies in consumption patterns. The diffusion of more knowledge and understanding of other regulatory regimes and experiences in planning elsewhere would help to finally dispel myths about a self-regulating continental style of drinking. Experiences from Britain and mainland Europe might usefully be transferred to developing countries who are likely to become the future targets of the major drinks operators.

Final comments

Sarah Thornton argues that 'the seemingly chaotic paths along which people move through the city are really remarkably routine' (Thornton, 1995: 91, cited in Chatterton and Hollands 2003: 5). The night-time economy has indeed developed in a fairly predictable pattern, with entertainment, socializing and the consumption of alcohol as important now as they have been in the past. Young men continue to dominate, as both producers and consumers, and, in some quarters, women continue to be cast as either an imposition into the traditional male city, as potential victims or as morally dubious. But there is another under-standing of the everyday and the 'routine' that has been used in this discus-sion, and that is the manner in which typically everyday activities such as working or shopping have shifted later into the night. We are far from a state where these activities are as common to the night as the day, and the night continues to be a time-space in which to relax rather than work. Nonetheless, the Fordist city where individuals worked 9 a.m.–5 p.m., and there was a sharp distinction between work and leisure, with leisure typically occurring only on Friday or Saturday evenings, is no longer the case.

In Chapter 2, citing Lovatt and O'Connor (1995: 131), it was argued that there remains a hidden array of stories yet to be told about the night-time city. In similar terms, Chatterton and Hollands have pointed to the impossibility of capturing the rich diversity of late-night culture. Far from the notion of a single 'Binge Britain', the night-time economy is as multifaceted and complex as the city during the day. As we have attempted to capture in this book, the city at night is marked by tensions, contradictions and competing interests. It is rational and sanitized on the one hand, emotionally driven and emotionally enhanced on the other (Chatterton and Hollands 2003: 4). It can be both rowdy and intimidating and entail dangers for some, but offer a realm of pleasurable and spontaneous opportunity for others. Beyond the chain bars and late-night disorder, an extensive variety of late-night cultures are occurring alongside, in opposition, or in complete ignorance of each other. Research has, in recent years, began to tap into this, and begun to chart the disparity of the late-night city, but the actual experiences and lives of many people at night continue to be drowned out by the more newsworthy tales of drug use, binge drinking and late-night violence.

Thornton is right to point to the routine and sometimes predictable paths that late-night producers and consumers follow. However, it is equally worth recalling Hubbard's point made earlier, that the routine or the everyday does not necessarily mean there is no room for change. The everyday, or the everynight in this instance, can also be the site for resistance and transformation. The simple act of walking one's dog through the streets at night challenges the often alarmist accounts of Binge Britain, where decent citizens are cast as fearful and afraid. Other activities or behaviours at night can also challenge the dominant commercialized world of late-night Britain; activities that are free, or creatively driven, for example, rather than motivated purely by profit.

Binge Britain remains the dominant framework for understanding the late-night city today. There are, however, a number of alternatives. These alternative frameworks operate as 'models' for how the late-night city could be, or as reminders of what it once was in other times and places. These models can be loosely phrased as 'the commercial model', 'the British nostalgic', 'European nostalgic', 'high-culture' and 'zoned'.

The dominant and most visible model for the night-time economy is the commercial model, which is held responsible for all manner of things from encouraging binge drinking to destroying the traditional British pub. For Terence Blacker, writing in the *Independent*, the demise of the local pub is even associated with increased suicides:

> Recently, there has been a debate in Ireland as to why levels of suicide and depression have been rising in recent years. The experts have generally agreed on one of the main causes. 'The pub was a social centre. It created a sense of togetherness', one of them said. 'That is all dying out now and people can find themselves alone, in some cases drinking alone, which can lead to an even greater sense of loneliness.'
>
> (Blacker 2008)

In the commercial model, profit, rather than 'access, equality or creativity' (Chatterton and Hollands 2003: 5) reigns. Of course, it is not simply 'profit' that causes alarm, but who makes that profit and the means by which they do so. Global or national companies, not tied to the local and driven by shareholders or faceless corporations, are of concern here.

In part, this model is logical because town centres are profit-making spaces. In a somewhat twisted logic, the late-night economies we now have make commercial sense. Bars, pubs and clubs that fail to turn over a profit quickly disappear, only to be replaced by venues that operate in terms closer to those that are more economically viable. There is also little room within current discourse for an alternative urban model where profit and commercialism fall far behind social, cultural or political motives.

In stark contrast to the commercial model, the 'nostalgic model' calls for a return to the British pub, a place of community, belonging and civility. British pubs are believed to have once been the centre of a local community, welcoming and accessible, and with bar-staff known by name. The commercialization of the alcohol industry has resulted in a reduction of such venues, replacing them with large, city centre, anonymous sites owned by distant interests. To borrow Marc Augé's term, these anonymous super pubs function as non-places (Augé 1995); these being sites lacking in an identifiable sense of identity, history or geography. In stark contrast to real places, pubs today are said to be soulless, homogeneous, aimed primarily at young drinkers and managed with a view simply to maximize profit, rather than sustaining or creating a sense of community.

There is much to be said for this account. In Chapter 9 it was noted that some consumers will go out of their way to visit country pubs, where it is possible to relax and dine in comfort. But, like all forms of nostalgia, this model reveals more about the present than the past. It highlights an ongoing anxiety about the breakdown of community and an unease with new forms of sociality and social behaviour. This model also ignores the fact that pubs, in the past and sometimes in the present, were never entirely welcoming or accessible to all. Equally, it presumes that contemporary chain pubs are little more than anonymous drinking establishments. Some irresponsible venues may encourage binge drinking and feel alienating for people over a particular age, but there is no evidence to suggest chain venues cannot also be welcoming and familiar spaces – for some segments of the community.

Related to the nostalgia model is the idealized vision of Europe which, as we have argued, served as a touchstone for discussion in the England and Wales Licensing Act 2003 and licensing reform in New South Wales, Australia. Whether it is referred to as café culture or a continental ambience, drinking and late-night culture on the continent are believed to be more civilized and marked by a convivial, child-friendly atmosphere. Like the British pub model, European late-night culture is understood as supporting local communities, with small intimate venues encouraging a more friendly form of relaxed interaction. According to the European model, sobriety is as much to do with individual restraint as a culture where respect, shame and community attitudes continue to actually mean something. However, as has been argued here, much of what can be seen in continental Europe owes to regulation rather than genetic predisposition. Problems with the night-time economy in Barcelona or the botellón phenomena, for example, also suggest that the European model is an ideal rather than a reality. For the European model to work in Britain requires more than just a cultural change, but regulation, control and a substantial re-ordering of the asset values of commercial property. To repeat a comment made by the Chief Constable of Nottinghamshire, 'if you want café culture, build cafés'.

The European model serves as an impetus for a related model that may be simply referred to as 'high culture'. This account calls for a greater diversity within the night-time economy by encouraging the growth of high cultural pursuits. Rather than simply bars and clubs aimed at young men and women, opening museums, galleries, theatres and late-night retail will, it is hoped, attract a greater diversity of patrons, particularly families. This in turn will serve to offer an alternative to a night in the pub, and drinkers may find a night at the theatre or a library a more rewarding way to spend their time. In planning terms, the high-culture model calls for greater mixed-use development and encouragement for other land-uses. It also requires councils to support and perhaps finance events, festivals and street markets. Consumers, as noted in Chapter 9, strongly support such interventions, which makes it all the more surprising that few councils have led the way in doing this.

However, like the other models explored here, there are limits to the high-culture model. First, it ignores that some people simply will not want to visit galleries or museums at night. Research conducted by the authors indeed found that nice bars and restaurants were a good deal more favoured than the pursuit of high-culture entertainment. The high-culture model also presumes that galleries, theatres, museums and the opera do not already exist alongside the commercial world. In smaller towns and cities alternatives to the commercial sector are thin, and more could be proposed. However, from our work, there is little evidence to suggest galleries and museums would be a popular way of spending an evening. Other family pursuits are more favourable, and art centres or interactive activities would be welcome. The model also presumes that consumers do not do both; as if it were not possible to enjoy the theatre on one night, followed by a drunken episode later on, or on another night. A final question about the high-culture model is its implications for spatial configuration, that is, would it require a separation of uses, with alcohol-centred activities on one street and festivals on another, or would it depend upon drinkers appearing later in the night, allowing others to venture out earlier in the evening?

On this note, there is the zoned model, which has been used elsewhere and, to a degree, in Britain. This is where different uses are separated from each other, such that family-orientated venues are zoned in one area and heavy drinking venues in another. This model already exists in some British towns and cities. Romford, for example, features a street dominated by youth-orientated clubs and bars, alongside a leisure centre featuring a cinema, bowling alley and family-style restaurants. Equipped with its own car park, the groups that patronize these contrasting entertainment venues in Romford do not have to necessarily cross paths. The zoned model goes against the current trend towards mixed communities, however. It also raises a series of questions about how we want the city at night to be, separated with distinctive uses for different groups, or a more integrated and therefore unpredictable, truly public space.

Each of these models has reach but, following Latham (2003), we need to think more productively about the city. Rather than starting from a question of what is wrong with the city at night, or noting how it does not fit within a predetermined model, he argues we need to think more critically about the 'spontaneous, loosely institutionalized, emergent trends' (Latham 2003: 1702). More provocatively, he calls on critics of the current urban form to explicitly define how diverse, inclusive, egalitarian, socially driven and demo-cratic urban forms may actually develop. How would they be achieved? How would they be sustained? And, 'what kind of compromises are they founded on?'. The challenge is to move beyond romantic visions of the city, and instead focus on the observed dynamic, spontaneous and under-represented relations that are occurring in the city today. This by no means suggests we should ignore the problems that currently exist in the late-night city. Instead, it calls on researchers and practitioners to examine and understand the less tidy and more unexplainable elements of the city at night. One such question could be, why do the vast majority of people not binge drink? Why, on the whole, and given the toxic mix of poor infrastructure, crowded venues and over consump-tion, is violence and anti-social disorder not more extensive?

Most challenging of all, researchers and policy-makers will need a more thorough engagement with the very notion of pleasure. The late-night city is a place of work and other everyday pursuits, but it remains, above all else, associated with the pursuit of entertainment, leisure and play. As we know, the issue at present is that some of this entertainment occurs at the expense of others. There are troubling trends within the night-time city that cannot be ignored. More legislation, by-laws and regulation could aid in minimizing this state of affairs and opening up the city to a greater level of diversity. Yet, as Talbot reminds us, the late-night city has always been caught between chaos and control, or regulation and freedom (Talbot 2006: 159). Tighter regulation of the industry has been advocated in this book, but we also need a more nuanced understanding of how the city at night does already operate, and second, the rec-ognition that some spaces of the city are already diverse and doing well. A more evidence-based and deeper understanding of the intersection between planning, the urban environment and the pursuit of pleasure is therefore necessary. Pleas-ure does not need to be associated exclusively with hedonism or excess.

This book has concluded with a discussion of the more everyday pleasures that people wished to have in the night-time city. These were places to sit and talk in comfort with friends, to eat out, to listen to live music, places to go out as a whole family and to enjoy the magical and spontaneous, as exemplified by a street festival or the experience of ice-skating. Above all, what was wished for was contact with other people. These are not new forms of solidarity, but reconfigurations of familiar activities, made possible by a rise in disposable incomes, some leisure time and technological advancement. It is a sad indictment of our approach to urban development in the UK that these simple pleasures are so difficult to achieve. We started by considering the term 'the night-time economy', and this is an appropriate point at which to end. Above all else, it is economic forces that have prevented the night-time city from developing as an inclusive space. Planning for the night-time city means just that, taking the economy out of the night-time economy and seeing the night-time city for what it is and what it might become.

Bibliography

10 Downing Street (2007) 'Press conference', Press Office. www.pm.gov.uk/output/Page12590. asp (accessed 23 July 2007).

Academy of Medical Sciences (2004) *Calling Time: The Nation's Drinking as a Major Health Issue*. London: Academy of Medical Sciences.

Ackroyd, P. (2001) *London: The Biography*. London: Vintage.

ACPO (2005) 'Memorandum submitted by the Association of Chief Police Officers of England, Wales and Northern Ireland', Home Affairs: Select Committee, House of Commons Session 2004–2005 HC 80-II. www.publications.parliament.uk/pa/cm200405/cmselect/cmhaff/80ii/80we03.htm (accessed 21 July 2008).

Ajuntament de Barcelona (2006) *Ordenanza de medidas para fomentar y garantizar la convivencia ciudadana en el espacio público de Barcelona B-11.503-2006*. Barcelona: Ajuntament de Barcelona.

Alavaikko, M. and Osterberg, E. (2000) The influence of economic interests on alcohol control policy: a case study from Finland, *Addiction* 95 (4): S565–79.

Alcock, A., Bentley, I., McGlynn, S. and Murrain, P. (1988) *Responsive Environments*. Oxford: Architectural Press.

Alexandroni, S. (2006) Stupid drinking, *New Statesman*, 27 November.

Alibhai-Brown, Y. (2005) The mixed messages that lost 'the war on drugs', *Evening Standard*, 6 July.

Allen, C. and Blandy, S. (2004) *City Centre Living: Implications for Urban Policy*. Sheffield: Centre for Economic and Social Research. www.communities.gov.uk/corporate/researchandstatistics/research1/crosscuttinganalyticalprogrammes/introduction/newhorizonsprogramme/worldclasstowns/future (accessed 17 July 2008).

Allen, J. and Pryke, M. (1994) The production of service space, *Environment and Planning D: Society and Space* 12 (4): 453–75.

Amin, A. and Thrift, N. (2002) *Cities: Reimaging the Urban*. Cambridge: Polity Press.

Anderson, P. and Baumberg, B. (2006) *Alcohol in Europe: A Public Health Perspective*. London: Institute of Alcohol Studies.

Anonymous (1999) It's a free for all in Temple Bar, *Irish Independent*, 13 July.

Anonymous (2004a) Aberdeen bid to raise drink prices foiled, *Caterer and Hotelkeeper*, 11 November. www.caterersearch.com/Articles/2004/11/11/56182/aberdeen-bid-to-raise-drink-prices-foiled.html (accessed 22 November 2008).

Anonymous (2004b) Sensible Drinking Supplement, *New Statesman*, 15 March.

Anonymous (2005) The price of a law nobody wants, *Daily Mail*, 15 January. www.dailymail.co.uk/pages/livearticles/news/newscomment/html?in_article_id+334223&in_pageid_1787 (accessed 24 November 2006).

Anonymous (2006) Wild things: Supplement, *Travel Weekly: The Choice of UK Travel Professionals*: 10–11.

Anonymous (2008) Venues: Liquid & Envy, Basildon, *Night*. www.nightmagazine.co.uk/venues/LiquidEnvyBasildon.htm (accessed 20 August 2008).

Antoniades, P., Maggi, B. and Pickering, E. (2007) *Alcohol, Entertainment and Late Night Refreshment Licensing*, Department for Culture, Media and Sport. www.culture.gov.uk/ Reference_library/rands/statistics/alcohol_entertainment_licensing_statistics.htm (accessed 17 November 2007).

Appleton, L. (1995) The gender regimes of American cities. In Garber, T. and Turner, R. (eds) *Gender in Urban Research*. Thousand Oaks: Sage.

Atkinson, R. (2003) Domestication by cappuccino or a revenge on urban space? Control and empowerment in the management of public spaces, *Urban Studies* 40 (9): 1829–43.

Augé, M. (1995) *Non-places: Introduction to an Anthropology of Supermodernity*. London: Verso.

Australian General Practice Network (2008) *Submission to the Senate Community Affairs Committee: Inquiry into Ready-to-Drink Alcohol Beverages*. Manuka, Australian General Practice Network.

Babb, P. (2007) *Violent Crime, Disorder and Criminal Damage Since the Introduction of the Licensing Act 2003*. London: The Home Office. www.homeoffice.gov.uk/rds/pdfs07/rdsolr 1607.pdf (accessed 16 July 2007).

Babor, T., Caetano, R., Casswell, S. and Edwards, G. (eds) (2003) *No Ordinary Commodity: Alcohol and Public Policy*. Oxford: Oxford University Press.

Balibrea, M.P. (2004) Urbanism, culture and the post-industrial city: challenging the 'Barcelona Model'. In Marshall, T. (ed.) *Transforming Barcelona*. London: Routledge, pp. 205–24.

Bannister, J., Fyfe, N. and Kearns, A. (2006) Respectable or respectful? (In)civility and the city, *Urban Studies* 43 (5/6): 919–37.

Barcelona Reporter (2006) Barcelona court sends noisy bar owner to prison for four years, *Barcelona Reporter/news and views from the Catalan capital*, 27 March. www.barcelonareporter. com/index.php/pg_print_article/barcelona_court_sends_noisy_bar_owner_to_prison_for_four_ years (accessed 26 November 2008).

Barcelona Turisme (2007) Tourism statistics in Barcelona 2006 (provisional). www.barcelonaturisme. com/?go=gNTyP9CiptZPIC4AStFCF834mptYKONXTNu3Luyy3gO7AYR9kOxyadtwYOQx4jwsFT 27knlfq9JMCN7n3LWqcBBAvevbboWnHNC6/j4KnMVGHA== (accessed 6 March 2008).

Barr, A. (1998) *Drink: A Social History*. London: Pimlico.

Bauman, Z. (2007) *Liquid Times: Living in an Age of Uncertainty*. Cambridge: Polity Press.

Bayliss, D. (2007) The rise of the creative city: culture and creativity in Copenhagen, *European Planning Studies* 15 (7): 889–903.

BBC News (2004) Binge drinking 'out of control', *BBC News Online*, 14 March. http://news.bbc. co.uk/1/hi/uk/3510084.stm (accessed 16 February 2004).

BBC News (2006) UK 'tops binge-drinking league', *BBC News Online*, 7 April. http://news.bbc. co.uk/1/hi/health/4886550.stm (accessed 20 October 2008).

BBC News (2007) Nude stag man's father apologises, *BBC News Online*, 7 June. http://news. bbc.co.uk/1/hi/uk/6731845.stm (accessed 15 May 2008).

BBC News (2008) Big increase in single households, *BBC News Online*, 8 May http://news.bbc. co.uk/1/hi/scotland/7390250.stm (accessed 4 September 2008).

BCN Living (2005) Noise in Gracia, *Barcelona Metropolitan*. www.barcelona-metropolitan.com/ Article (accessed 6 March 2008).

Beck, U. (1992) *Risk Society: Towards a New Modernity*. London: Sage.

Beck, U. and Beck-Gernsheim, E. (2001) *Individualization: Institutionalized Individualism and Its Social and Political Consequences*. London: Sage.

Bell, D. (2006) Commensality, urbanity, hospitality. In Lashley, C., Lynch, P. and Morrison, A. (eds) *Critical Hospitality Studies*. London: Butterworth Heinemann.

Bell, D. (2007) The hospitable city: social relations in commercial spaces, *Progress in Human Geography* 31 (7): 7–22.

Bell, D. (2008) The views of outside in British city centres, *Urban Design* 108: 24–7.

Bell, D. and Binnie, J. (2005) What's eating Manchester? Gastro-culture and urban regneration, *Architectural Design* 75 (3): 78–85.

Bibliography

Bell, D. and Jayne, M. (eds) (2004) *City of Quarters*. Aldershot: Ashgate.

Bennett, J. and Dixon, M. (2006) *Single Person Households and Social Policy: Looking Forwards*. York: Joseph Rowntree Foundation.

Berman, M. (1983) *All That is Solid Melts into Air: The Experience of Modernity*. London: Verso.

BERR (2008) *Competition Law: Issues Which Arise for Business when the Government Seek to Encourage Businesses to Work Together to Deliver Desired Policy Outcomes*. London, Department for Business Enterprise and Regulatory Reform. www.berr.gov.uk/files/file45711. pdf (accessed 25 July 2008).

Berridge, V., Thom, B. and Herring, R. (2007) *The Normalisation of Binge Drinking? An Historical and Cross Cultural Investigation with Implications for Action*, Alcohol Education and Research Council. www.aerc.org.uk/documents/pdfs/finalReports/AERC_FinalReport_0037.pdf (accessed 12 February 2008).

Better Regulation Task Force (1998) *Licensing Legislation*, London Central Office of Information.

Bianchini, F. (1995) Night cultures, night economies, *Planning Practice and Research* 10 (2): 121–6.

Bianchini, F. and Ghilardi, L. (2004) The culture of neighbourhoods: a European perspective. In Bell, D. and Jayne, M. (eds) *City of Quarters*. Aldershot: Ashgate, pp. 237–48.

Blacker, T. (2008) When pubs die, we are all poorer, *Independent*, 20 February. www.independent.co.uk/opinion/commentators/terence-blacker/terence-blacker-when-pubs-die-we-are-all-the-poorer-784367.html (accessed 20 February 2008).

Blair, T. (2005) Commons Hansard column 300, 12 January.

Blau, R. (2008) Checkout any time you like . . ., *Financial Times Weekend*, 9/10: 14.

Boella, M.J., Legrand, W., Pagnon-Maudet, C., Sloan, P. and Baumann, A. (2005) Regulation of the sale and consumption of alcoholic drinks in France, England and Germany, *International Journal of Hospitality Management* 25: 398–413.

Borsay, P. (2007) Binge drinking and moral panics: historical parallels?, *History and Policy*. www.historyandpolicy.org/papers/policy-paper-62.html#parallels (accessed 28 November 2007).

Bourdieu, P. (1992) *Distinction: A Social Critique of the Judgement of Taste*. London: Routledge and Kegan Paul.

Brannen, J. and Nilsen, A. (2005) Individualisation, choice and structure: a discussion of current trends in sociological analysis, *The Sociological Review* 53 (4): 799–803.

British Beer and Pub Association (2008) *Reference: Industry Statistics: Tax and Expenditure*. www.beerandpub.com/industryArticle.aspx?articleId=129 (accessed 9 April 2008).

Bromley, R., Thomas, C. and Millie, A. (2001) Exploring safety concerns in the night-time city, *Town Planning Review* 71 (1): 71–96.

Bromley, R.D.F., Tallon, A.R. and Thomas, C.J. (2003) Disaggregating the space–time layers of city centre activities and their users, *Environment and Planning A* 35: 1831–51.

Bromley, R.D.F., Tallon, A.R. and Thomas, C.J. (2005) City centre regeneration though residential development: contributing to sustainability, *Urban Studies* 42 (13): 2407–29.

Budd, T. (2003) *Alcohol-related Assault: Findings from the British Crime Survey*, London: Home Office.

Burnett, J. (1999) *Liquid Pleasures: A Social History of Drinks in Modern Britain*. London: Routledge.

Burke, T. (1943) *English Night-Life: From Norman Curfew to Present Black-Out*. London: BT Batsford.

Butler, T. (2003) *London Calling: The Middle Classes and the Re-making of Inner London*. Oxford: Berg.

Cabinet Office (2004) *National Alcohol Harm Reduction Strategy for England*. London: The Stationery Office.

Campo, D. and Ryan, B. (2008) The entertainment zone: unplanned nightlife and the revitalization of the American downtown, *Journal of Urban Design* 13 (3): 291–315.

Carmona, M. (2004) *Living Places: Caring for Quality, Managing External Public Space*. London: Office of the Deputy Prime Minister. www.communities.gov.uk/publications/communities/livingplacescaring (accessed 22 July 2008).

Castells, M. (2000) *The Rise of the Network Society*. Oxford: Blackwell.

Central Office for Information (1998) *Better Regulation Task Force Licensing Legislation*. London: Central Office for Information.

Chatterton, P. and Hollands, R. (2002) Theorising urban playscapes: producing, regulating and consuming youthful nightlife city spaces, *Urban Studies* 39 (1): 95–116.

Chatterton, P. and Hollands, R. (2003) *Urban Nightscapes: Youth Cultures, Pleasure Spaces and Corporate Power*. London: Routledge.

Chatterton, P., Hollands, R., Byrnes, B., Read, C. and Aubrey, M. (2002) *Youth Culture, Nightlife and Urban Change*. Newcastle-upon-Tyne: University of Newcastle-upon-Tyne. www.ncl.ac.uk/youthnightlife/home.htm (accessed 10 February 2008).

Cheltenham Borough Council (2004) *Cheltenham's Night-time Economy Strategy 2004–2007*. Cheltenham: Cheltenham Borough Council.

City of Melbourne and Gehl Architects (2005) *Places for People: Melbourne 2004*. City of Melbourne: Melbourne.

City of Norwich (2004) *Replacement Local Plan*. Norwich: Department of Planning.

City of Westminster (2008) *History of Waste Management*. www.westminster.gov.uk/environment/rubbishwasteandrecycling/rubbishandwaste/householdwaste/history.cfm.

City Planning Group (2006) *Entertainment Supplementary Planning Guidance*. London: City of Westminster.

Civic Trust (2005) *Case Study: Manchester City Centre Safe*. London: The Civic Trust. www.civictrust.org.uk/evening/studies.shtml (accessed 13 June 2008).

Clark, T.J. (1999) *The Painting of Modern Life: Paris in the Art of Manet and his Followers*. London: Thames and Hudson.

Clarke, T.N. (2004) *The City as an Entertainment Machine*. Oxford: Elsevier.

Clennell, A. (2007) 'Pubs filthy over Labor rebuff on small bars', *Sydney Morning Herald*, 7 November: 4.

Cohen, S. (1972) *Folk Devils and Moral Panics: The Creation of the Mods and Rockers*. Oxford: Martin Robinson.

Coleman, R. (2004) Watching the degenerate: street camera surveillance and urban regeneration, *Local Economy* 19 (3): 199–211.

Collins, J. (2004) *Issues Around the Evening Economy Conference*. London: Civic Trust. www.civictrust.org.uk/evening/events.shtml.

Comedia (1991) *Out of Hours: A Report for the Gulbenkian Foundation*. London: Demos and Comedia.

Conisbee, M., Kjell, P., Oram, J., Palmer, J.B., Simms, A. and Taylor, J. (2004) *Clone Town Britain: The Loss of Local Identity on the Nation's High Streets*. London: New Economics Foundation. www.neweconomics.org/gen/z_sys_publicationdetail.aspx?pid=189 (accessed 15 November 2008).

Connolly, P. (2005) Binge drinking and Leeds 'the 24 hour "European" city', *Leeds Civic Trust Newsletter*.

Cook, I.R. (2008) Mobilising urban policies: the policy transfer of US business improvement districts to England and Wales, *Urban Studies* 45 (4): 773–95.

Cottam, N. (2006) Working overtime, *Marketing*, 21 January.

Coupland, A. (ed.) (1997) *Reclaiming the City: Mixed Use Development*. London: E & FN Spon.

Cozens, P.M., Neale, R.H., Whitaker, J., Hillier, D. and Graham, M. (2003) A critical review of street lighting, crime and fear of crime in the British city, *Crime Prevention and Community Safety: An International Journal* 5: 7–24.

Creagh, S. (2007) Hotel warlords rile Keating, *Sydney Morning Herald*, 27 September. www.

smh.com.au/news/national/hotel-warlords-rile-keating/2007/09/26/1190486395849.html (accessed 28 September 2008).

Cruikshank, D. and Burton, N. (1990) *Life in the Georgian City*. London: Viking.

Davies, P. and Mummery, H. (2006) *NightVision: Town Centres For All*. London: The Civic Trust.

Day, K., Gough, B. and McFadden, M. (2004) Warning! Alcohol can seriously damage your feminine health, *Feminist Media Studies* 4 (2): 165–83.

DCLG (2005a) *Planning Policy Statement 1: Delivering Sustainable Communities*. London: The Stationery Office.

DCLG (2005b) *Planning Policy Statement 6: Planning for Town Centres*. London: The Stationery Office.

DCLG (2006a) *Design Coding in Practice: An Evaluation*. Wetherby: Department of Communities and Local Government. www.communities.gov.uk/publications/citiesandregions/designcoding2 (accessed 28 July 2008).

DCLG (2006b) *Good Practice Guide on Planning for Tourism*. London: Government, Department for Communities and Local Government. www.communities.gov.uk/documents/planningandbuilding/pdf/151753.pdf (accessed 15 November 2008).

DCLG (2008) *Proposed Changes to Planning Policy Statement 6: Planning for Town Centres Consultation*, London: Department for Communities and Local Government.

Deehan, A. (1999) *Alcohol and Crime: Taking Stock*. London: Home Office.

Degen, M. (2003) Fighting for the global catwalk: formalizing public life in Castelfield (Manchester) and diluting public life in el Raval (Barcelona), *International Journal of Urban and Regional Research* 27 (4): 867–80.

Degen, M. (2004) Barcelona's games: the Olympics, urban design and global tourism. In Urry, J. and Sheller, M. (eds) *Global Places to Play*. London: Routledge, pp. 131–42.

Degen, M. (2008) *Sensing Cities: Regenerating Public Life in Barcelona and Manchester*. London: Routledge.

Department for Children, Schools and Families, The Home Office and Department of Health (2008) *Youth Alcohol Action Plan*. London: The Stationery Office.

Department for Culture, Media and Sport (2003a) *Major Reform of Licensing Laws Completed*. www.culture.gov.uk/reference_library/media_releases/2689.aspx.

Department of Culture, Media and Sport (2003b) *Regulatory Impact Assessment: Licensing Bill*.

Department for Culture, Media and Sport (2007) *Number of Licensed Premises Revealed*. www.culture.gov.uk/Reference_library/Press_notices/archive_2007/dcms133_07.htm?contextId= {C59B4287–7256–46DE-A310–2F11E65971C9 (accessed 16 November 2007).

Department for Culture, Media and Sport, The Home Office and Office of the Deputy Prime Minister (2005) *Drinking Responsibly: The Government's Proposals*. London: Department for Culture, Media and Sport.

Department of Health and Human Services (2004) NIAAA Council approves definition of binge drinking, *NIAAA Newsletter*.

Department of the Environment (1996) *Planning Policy Guidance Note 6: Planning for Town Centres*. London: The Stationery Office.

Dispatches (2008) *The Hidden World of Lap Dancing*, Channel 4.

Dooldeniya, M.D., Khafagy, R., Mashaly, H., Browning, A.J., Sundaram, S.K. and Biyani, C.S. (2007) Lower abdominal pain in women after binge drinking, *British Medical Journal* 335: 992–3.

Douglas, M. (ed.) (1987) *Constructive Drinking: Perspectives on Drinking from Anthropology*. Cambridge: Cambridge University Press.

Drummond, D.C. (2004) An alcohol strategy for England: the good, the bad and the ugly, *Alcohol and Alcoholism* 39 (5): 377–9.

Eden, I. (2007) *Inappropriate Behaviour: Adult Venues and Licensing in London*. London: The

Lilith Project. www.object.org.uk/downloads/Inappropriate_Behaviour.pdf (accessed 29 October 2008).

Eldridge, A. and Roberts, M. (2008) 'A comfortable night out?': alcohol, drunkenness and inclusive town centres, *Area* 40 (3): 365–74.

Eltham, B. (2008) Land of the rum rebellion, *New Matilda*. http://newmatilda.com/2008/05/08/rum-rebellion (accessed 8 May 2008).

Elvins, M. and Hadfield, P. (2003) *West End 'Stress Area', Night-time Economy Profiling: A Demonstration Project*. Durham: University of Durham.

Energy Control (2007) *Safer Nightlife Projects: A European Proposition to Promote Safer Nightlife and Share Good Practice*. Asociacion Bienestar y Desarollo-Spain, Barcelona, DC & DI Safer Nightlife Group.

ESRC Society Now (2008) Radical re-think required on drinking policies, *Society Now* 1.

Esteban, J. (2004) The planning project: bringing value to the periphery, recovering the centre. In Marshall, T. (ed.) *Transforming Barcelona*. London: Routledge, pp. 111–50.

Evans, G. (2001) *Cultural Planning: An Urban Renaissance?* London: Routledge.

Featherstone, M. (1991) *Consumer Culture and Postmodernism*. London: Sage.

FECALON (2007) Las Actas de la Guardia Urbanan avalan que los locales nocturnos son poce ruidosos, *FECALON*, 6 April.

Florida, R. (2002) *The Rise of the Creative Class: And How It is Transforming Work, Leisure, Community and Everyday Life*. New York: Basic Books.

FM-News (1999) *News Extracts*.

Foucault, M. (1986) Of other spaces, *Diacritics* 16 (Spring): 22–7.

Fowler, R. (1991) *Language in the News: Discourse and Ideology in the Press*. London: Routledge.

Fox, K. (2005) *Watching the English: The Hidden Rules of English Behaviour*. London: Hodder.

Fox, K. and Marsh, P. (2006) *Social and Cultural Aspects of Drinking: A Report to the Amsterdam Group*. Oxford: Social Issues Research Centre.

Fraser, C. (2007) Attitudes to alcohol in Europe, *BBC News/UK*, 13 November. http://news.bbc.co.uk/1/low/uk/7093143.stm (accessed 28 November 2007).

Garcia, M. and Claver, N. (2003) Barcelona: governing coalitions, visitors, and the changing city center. In Hoffman, L.M., Fainstein, S.S. and Judd, D.R. (eds) *Cities and Visitors: Regulating People, Markets and City Space*. Oxford: Blackwell, pp. 113–25.

Garcia-Ramon, M.D. and Albet, A. (2000) Commentary, *Environment and Planning A* 32: 1331–4.

Gask, A. (2004) *Anti-Social Behaviour Orders and Human Rights*. London: Liberty. www.liberty-human-rights.org.uk/issues/pdfs/asbos-and-human-rights.pdf (accessed 22 July 2008).

Gauntlett, D. (1998) Ten things wrong with the 'effects model'. In Harindranath, R. and Linné, O. (eds) *Approaches to Audiences: A Reader*. London: Arnold.

Gehl, J. and Gemzøe, L. (2001) *New City Spaces*. Copenhagen: Danish Architectural Press.

Gehl, J., Gemzøe, L., Kirknaes, S. and Sternhagen, B.S. (2006) *New City Life*. Copenhagen: Danish Architectural Press.

Geiger, S. (2007) Exploring night-time grocery shopping behaviour, *Journal of Retailing and Consumer Services* 14 (1): 24–34.

Giddens, A. (1998) *The Third Way: The Renewal of Social Democracy*. Cambridge: Polity.

Gill, M. and Spriggs, A. (2005) *Assessing the Impact of CCTV*. London: The Home Office. www.homeoffice.gov.uk/rds/pdfs05/hors292.pdf (accessed 17 July 2008).

GLA (2007) *Managing the Night Time Economy: Best Practice Guidance*. London: Greater London Authority.

GLA (2008) *Information*. London: Greater London Authority. www.london.govuk+Night+buses+expansion+2000-2007+London+Transport+for+London.

GLA Economics (2003) *Spending Time: London's Leisure Economy*. London: Greater London Authority.

Bibliography

Glancey, J. (2008) Why this year's Stirling prize is cooler than ever. http://blogs.guardian.co.uk/art (accessed 18 July 2008).

Glorieux, I., Mestdag, I. and Minnen, J. (2008) The coming of the 24-hour economy? Changing work schedules in Belgium between 1966 and 1999, *Time & Society* 17 (1): 63–83.

Goodall, T. and Lawrence, K. (2007) *Leeds Alcohol Strategy 2007–2010*. Leeds: Leeds Primary Care Trust. www.leedsinitiative.org/safer/page.aspx?id=2446 (accessed 9 April 2008).

Graham, K. and Homel, R. (2008) *Raising the Bar: Preventing Aggression in and Around Bars, Pubs and Clubs*. Cullompton: Willan Publishing.

Graham, S. and Marvin, S. (1996) *Telecommunications and the City: Electronic Spaces, Urban Places*. London: Routledge.

Grazian, D. (2007) *On the Make: The Hustle of Urban Nightlife*. Chicago: University of Chicago Press.

Greed, C. (2003) *Inclusive Urban Design: Public Toilets*. London: Architectural Press.

Greenacre, S. and Brown, J. (n.d.) *City Centre Safe*. Greater Manchester Police: City Centre Safe Unit. www.crimereduction.homeoffice.gov.uk/gp/gpvca01c.ppt (accessed 15 March 2009).

Griffiths, M. (2003) *Town Centre CCTV: An Examination of Crime Reduction in Gillingham, Kent*. London: The Home Office.

Gual, A. (2006) Alcohol in Spain: is it different?, *Addiction* 101: 1073–7.

Guise, J.M.F. and Gill, J.S. (2007) 'Binge drinking? It's good, it's harmless fun': a discourse analysis of accounts of female undergraduate drinking in Scotland, *Health Education Research* 22 (6): 895–906.

Gunter, B., Hansen, A. and Touri, M. (2008) *The Representation and Reception of Meaning in Alcohol Advertising and Young People's Drinking*. University of Leicester, Department of Media and Communication. www.aerc.org.uk/documents/pdfs/finalReports/AERC_FinalReport_0044.pdf.

Hackney Council (2007) *Housing in Hackney: Hackney Borough Profile 2006*. London: London Borough of Hackney.

Hadfield, P. (2004) Invited to binge? (licensing and the 24 hour city), *Town and Country Planning*, 1 July 73 (7/8): 235.

Hadfield, P. (2006) *Bar Wars: Contesting the Night in Contemporary British Cities*. Oxford: Oxford University Press.

Hadfield, P. (2007) Party invitations: New Labour and the (de)regulation of pleasure, *Criminal Justice Matters* 67 (1): 18–47.

Hadfield, P. and Traynor, P. (2008) *Policing and Regulation of the Night-time Economy: Recent Developments in Law and Practice*. ESRC Seminar: the Governance of Anti-Social Behaviour in Urban Spaces in the Night-time Economy, Leeds.

Hamnett, C. (2002) *Unequal City: London in the Global Arena*. London: Routledge.

Hannigan, J. (1998) *Fantasy City: Pleasure and Profit in the Postmodern Metropolis*. London: Routledge.

Harvey, D. (1985) *Consciousness and the Urban Experience*. Oxford: Blackwell.

Harvey, D. (2004) The right to the city. In Lees, L. (ed.) *The Emancipatory City: Paradoxes and Possibilities*. London: Sage, pp. 236–9.

Hayward, K. and Hobbs, D. (2007) Beyond the binge in 'booze Britain': market-led liminalization and the spectacle of binge drinking, *British Journal of Sociology* 58 (3): 437–55.

Heath, T. (1997) The twenty-four hour city concept – a review of initiatives in British cities, *Journal of Urban Design* 2 (2): 193–204.

Heather, N. (2006) Britain's alcohol problem and what the UK government is (and is not) doing about it, *Adicciones* 18 (3): 225–36.

HM Government (2007) *Safe. Sensible. Social. The next steps in the National Alcohol Strategy*. London: Department of Health and Home Office. www.dh.gov.uk/en/Publicationsandstatistics/Publications/PublicationsPolicyAndGuidance/DH_075218 (accessed 25 July 2008).

Hobbs, D. (2002) Mayhem after midnight in party town, *Sunday Times*.

Hobbs, D. (2005a) 'Binge drinkers': Folk devils of the binge economy. www.esrcsocietytoday. ac.uk/ESRCInfoCentre/about/CI/CP/research_publications/seven_sins/gluttony/nighttime.aspx ?ComponentId=10894&SourcePageId=11003.

Hobbs, D. (2005b) *Gluttony: 'Binge Drinking' and the Binge Economy*. Economic and Social Research Council. www.esrc.ac.uk/ESRCInfoCentre/about/CI/CP/research_publications/seven_ sins/gluttony/index.aspx?ComponentId=10890&SourcePageId=11077.

Hobbs, D., Hadfield, P., Lister, S. and Winlow, S. (2001) The '24-hour city' – condition critical?, *Town and Country Planning* 70 (11): 300–2.

Hobbs, D., Hadfield, P., Lister, S. and Winlow, S. (2003) *Bouncers: Violence and Governance in the Night-time Economy*. Oxford: Oxford University Press.

Hobbs, D., Lister, S., Hadfield, P., Winlow, S. and Hall, S. (2000) Receiving shadows: governance and liminality in the night-time economy, *British Journal of Sociology* 51 (4): 701–17.

Hobbs, D., O'Brien, K. and Westmarland, L. (2007) Connecting the gendered door: women, violence and doorwork, *British Journal of Sociology* 58 (1): 21–38.

Hobbs, D., Winlow, S., Hadfield, P. and Lister, S. (2005) Violent hypocrisy: governance and the night-time economy, *European Journal of Criminology* 2 (2): 161–83.

Hollands, R. and Chatterton, P. (2003) Producing nightlife in the new urban entertainment economy: corporatization, branding and market segmentation, *International Journal of Urban and Regional Research* 27 (2): 361–85.

Holloway, S., Jayne, M. and Valentine, G. (2008) 'Sainsbury's is my local': English alcohol policy, domestic drinking practices and the meaning of home, *Transactions of the Institute of British Geographers* 33 (4): 532–47.

Home Office (2001) *Time for Reform: Proposals for the Modernisation of our Licensing Laws*. www.culture.gov.uk/Reference_library/Publications/archive_2001 (accessed 11 July 2008).

Home Office (2007a) *Designation Orders: Best Practice*. www.crimereduction.homeoffice.gov. uk/alcoholorders/alcoholorders07.htm (accessed 13 June 2008).

Home Office (2007b) *Designation Orders*. www.crimereduction.homeoffice.gov.uk/alcoholor- ders/alcoholorders07.htm#Man

Homes, A.M. (2007) *Finalised Edinburgh City Local Plan Written Statement*. City Development, the City of Edinburgh Council.

Hoskins, G. and Tallon, A. (2004) Promoting the 'urban idyll': policies for city centre living. In Johnstone, C. and Whitehead, M. (eds) *New Horizons in British Urban Policy: Perspectives on New Labour's Urban Renaissance*. Aldershot: Ashgate.

Houlbrook, M. (2006) *Queer London: Perils and Pleasures in the Sexual Metropolis, 1918–1957*. London: University of Chicago Press.

House of Commons (2003a) *The Evening Economy and the Urban Renaissance*. ODPM: Housing Planning Local Government and the Regions Committee. London: The Stationery Office.

House of Commons (2003b) *The Evening Economy and the Urban Renaissance: Memoranda Submitted to the Urban Affairs Sub-committee*. ODPM: Housing Planning Local Government and the Regions Committee. London: The Stationery Office.

House of Commons (2004) *Minutes of Evidence taken before Home Affairs Committee on Anti- Social Behaviour: Uncorrected Transcript*. London: The Stationery Office.

House of Commons (2004/5) *Memorandum by the Bolton Townsafe Partnership Group (EVN 03)*. ODPM: Housing Planning Local Government and the Regions, House of Commons Cm 200405. www.publications.parliament.uk/pa/cm200405/cmselect/cmodpm/456/456we01.htm (accessed 25 July 2007).

House of Commons (2005) *Anti-Social Behaviour: Minutes of Evidence*. Select Committee on Home Affairs, HC 80-III. www.publications.parliament.uk/pa/cm200405/cmselect/cmhaff/ 80/4122101.htm (21 August 2008).

House of Commons (2006) *High Street Britain: 2015*. National Federation of Subpost Masters,

Bibliography

Home Office. www.nfsp.org.uk/uploads/pdfs/High%20Street%20Britain%202015%20report. pdf(accessed 22 November 2008).

House of Commons (2008) *Hansard Debates Westminster Hall 24 June Column 54WH*. London: The Stationery Office.

Howard-Pitney, B., Johnson, M.D., Altman, D.G., Hopkins, R. and Hammond, N. (1991) responsible alcohol service: a study of server, manager, and environmental impact, *American Journal of Public Health* 81 (2): 197–9.

Hubbard, P. (2005) The geographies of 'going out': emotion and embodiment in the evening economy. In Davidson, J., Bondi, L. and Smith, M.A. (eds) *Emotional Geographies*. Aldershot: Ashgate.

Hubbard, P. (2006) *City*. London: Routledge.

Hughes, K., Anderson, Z., Morleo, M. and Bellis, M.A. (2007) Alcohol, nightlife and violence: the relative contributions of drinking before and during nights out to negative health and criminal justice outcomes, *Addiction* 103 (1): 60–5.

Hunt, A. and Manchester, C. (2007) The Licensing Act and its implementation: nanny knows the 'third' way is best, *Web Journal of Current Legal Issues*. www.webjcli.ncl.ac.uk/2007/issue1/ rtf/hunt1.rtf (accessed 27 November 2008).

Hunter, G. (2007) Unruly Celtic fans face crackdown in Barcelona, *The Sunday Herald*, 23 December. http://findarticles.com/p/articles/mi_qn4156/is_/ai_n21172519 (accessed 27 November 2008).

Hutton, F. (2006) *Risky Pleasures? Club Cultures and Feminine Identities*. Aldershot: Ashgate.

Institute of Alcohol Studies (2007) *IAS Factsheet: Binge Drinking*. London: Institute of Alcohol Studies. www.ias.org.uk/resources/factsheets/binge_drinking.pdf (accessed 8 February 2008).

Institute of Alcohol Studies (July 2007) *Crime & Disorder: Binge drinking and the Licensing Act*. Cambridge: Institute of Alcohol Studies. www.ias.org.uk/resources/papers/crime_disorder.pdf (accessed 2 March 2006).

InsureandGo (2006) *Hen and Stag Revellers*. http://press.insureandgo.com/PressReleases. asp?pid=135 (accessed 21 February 2007).

Jackson, C. and Tinkler, P. (2007) 'Ladettes' and 'modern girls': 'troublesome' young femininities, *Sociological Review* 55 (2): 251–72.

Jackson, P. and Thrift, N. (1995) Geographies of consumption. In Miller, D. (ed.) *Acknowledging Consumption: A Review of New Studies*. London: Routledge.

Jacobs, J. (1961) *The Death and Life of Great American Cities: the Failure of Modern Town Planning*. New York: Random House.

Jameson, F. (1991) *Postmodernism, or, The Cultural Logic of Late Capitalism*. London: Verso.

Järvinen, M. and Room, R. (eds) (2007) *Youth Drinking Cultures: European Experiences*. Aldershot: Ashgate.

Jayne, M. (2006) *Cities and Consumption*. Abingdon: Routledge.

Jayne, M., Holloway, S.L. and Valentine, G.S. (2006) Drunk and disorderly: alcohol, urban life and public space, *Progress in Human Geography* 30 (4): 451–68.

Jayne, M., Valentine, G. and Holloway, S. (2008) Geographies of alcohol, drinking and drunkenness: a review of progress, *Progress in Human Geography* 32 (2): 247–63.

Jolly, I. (1994) *Policing the City at Night: Eyes on the Street*. The 24-Hour City: Selected Papers from the First National Conference on the Night-time Economy, Manchester.

Kelling, G.W. and Wilson, J.Q. (1982) Broken windows. www.theatlantic.com/doc/198203/broken-windows (accessed 24 July 2007).

Kershaw, C., Budd, T., Kinshott, G., Mattinson, J., Mathew, P. and Myhill, A. (2000) *The British Crime Survey England and Wales 2000*. London: Home Office.

Kettle, M. (2003) Alcoholic Britain should not be offered another drink, *Guardian*, 2 January: 14.

Kneale, J. (1999) 'A problem of supervision': moral geographies of the nineteenth-century British public house, *Journal of Historical Geography* 25 (3): 333–48.

KPMG llp (2008) *Review of the Social Responsibility Standards for the Production and Sale of Alcoholic Drinks: Volume 1*. London: The Home Office. http://drugs.homeoffice.gov.uk/publication-search/alcohol/alcohol-industry-responsibility (accessed 27 November 2008).

Kreitzman, L. (1999) *The 24 Hour Society*. London: Profile Books.

Kunzmann, K.R. (2004) Culture, creativity and spatial planning, *Town Planning Review* 75 (4): 383–404.

Lancashire Constabulary: Western Division (2004) *Nightsafe: Alcohol Related Violence Reduction Initiative*. Blackpool: Lancashire Constabulary.

Lang, E., Stockwell, T., Rydon, P. and Beel, A. (1998) Can training bar staff in responsible practices reduce alcohol-related harm?, *Drug and Alcohol Review* 17 (1): 39–50.

Lange, B. (2006) From cool Britannia to generation Berlin? Geographies of culturepreneurs and their creative milieus in Berlin. In Eisenberg, C., Gerlach, R. and Handke, C. (eds) *Cultural Industries: The British Experience in International Perspective*. Berlin: Humboldt University.

Lash, S. and Urry, J. (1994) *Economies of Signs and Space*. London: Sage.

Latham, A. (2003) Urbanity, lifestyle and making sense of the new cultural economy: notes from Auckland, New Zealand, *Urban Studies* 40 (9): 1699–724.

Lefebvre, H. (1991) *The Production of Space*. Oxford: Blackwell.

Levy, A. and Scott-Clark, C. (2004) Under the influence, *Guardian*, 20 November: 14.

Lewis, J. and Bennett, F. (2003) Themed issue on gender and individualisation, *Social Policy and Society* 3 (1): 43–5.

Ley, D. (1996) *The New Middle Class and the Remaking of the Central City*. Oxford and New York: Oxford University Press.

Light, R. (2005) The Licensing Act 2003: liberal constraint?, *Modern Law Review* 68 (2): 268–85.

Lindsay, J. (2005) *Drinking in Melbourne Pubs and Clubs: A Study of Alcohol Consumption Contexts*. Melbourne: School of Political and Social Inquiry, Monash University.

Local Government Association (5 June 2008) One in three councils considering Alcohol Disorder Zones. www.lga.gov.uk/lga/core/page.do?pageId=672652.

Lovatt, A. (1994) *Cultural Identity Through the Night-time Economy*. The 24-Hour City: Selected Papers from the First National Conference on the Night-time Economy, Manchester.

Lovatt, A. (1996) The ecstasy of urban regeneration: regulation of the night-time economy in the transition to a post-Fordist city. In O'Connor, J. and Wynne, D. (eds) *From the Margins to the Centre: Cultural Production and Consumption in the Post-industrial City*. Aldershot: Ashgate.

Lovatt, A. and O'Connor, J. (1995) Cities and the night-time economy, *Planning Practice and Research* 10 (2): 127–33.

Lovatt, A., O'Connor, J., Montgomery, J. and Owens, P. (1994) The 24-Hour City: Selected Papers from the First National Conference on the Night-time Economy, Manchester.

Lucas, K. (2004) Locating transport as a social policy problem. In Lucas, K. (ed.) *Running On Empty: Transport, Social Exclusion and Environmental Justice*. Bristol: Policy Press, pp. 7–14.

Luminar plc (2007) Annual Report. www.luminar.co.uk/media/upload_files/AnnualReport2007.pdf (accessed 21 July 2008).

McCarthy, J. (2005) Cultural quarters and regeneration: the case of Wolverhampton, *Planning Practice and Research* 20 (3): 297–311.

McDonald, F. (1998) Weekend influx of lager louts causes concern, *Irish Times*, 22 August.

McDonald, F. (2000) Work in Progress, *Irish Times*, 5 August.

McNeill, A. (2001) Still time for reform, *Alcohol Alert* 3. www.ias.org.uk/resources/publications/alcoholalert/alert200103/al200103_index.html (accessed 15 November 2007).

Marsh, P. and Kibby, K. (1992) *Drinking and Public Disorder*. London: The Portman Group. www.sirc.org/publik/dandpd.pdf (accessed 11 July 2008).

Marshall, S. (2005) *Streets and Patterns*. London: Spon.

Marshall, T. (ed.) (2004) *Transforming Barcelona*. London: Routledge.

Bibliography

Marx, K. (1977) *Capital: A Critique of Political Economy Vol. 1, Book One: The Process of Production of Capital.* London: Lawrence and Wishart.

Massey, D. (1998) *Space, Place and Gender.* Cambridge: Polity Press.

Mayor of London (2004) *A Managed Approach to the Night Time Economy in London: Research Study.* London: Greater London Authority. www.london.gov.uk/mayor/strategies/sds/camden-town.jsp (accessed 25 Juy 2008).

Mayor of London (2007) *Safer Travel at Night.* www.london.gov.uk/mayor/safer_travel (accessed 24 July 2008).

MCM Research (2003) *Implications for Noise Disturbance Arising from the Liberalisation of Licensing Laws: Report of Research and Consultation Conducted by MCM Research Ltd for the Department for Environment, Food and Rural Affairs (Defra).* London: MCM Research Ltd.

Measham, F. (2004) The decline of ecstasy, the rise of 'binge' drinking and the persistence of pleasure, *Probation Journal* 51 (4): 309–26.

Measham, F. (2006) The new policy mix: alcohol, harm minimisation, and determined drunkenness in contemporary society, *International Journal of Drug Policy* 17: 258–68.

Measham, F. and Brain, K. (2005) 'Binge' drinking, British alcohol policy and the new culture of intoxication, *Crime, Media, Culture* 1 (3): 262–83.

Melbin, M. (1978) Night as frontier, *American Sociological Review* 43 (1): 3–22.

Metropolitan Police Clubs and Vice Operational Command Unit (2003) *Preliminary Assessment of the Impact of the Licensing Act 2003 on the Metropolitan Police Service.* London: Metropolitan Police Service. www.met.police.uk/foi/pdfs/other_information/units/licensing_act_2003_impact.pdf (accessed 19 November 2007).

Midanik, L.T. (2006) Perspectives on the validity of self-reported alcohol use, *Addiction* 84 (12): 1419–23.

Miles, M. (2005) Interruptions: testing the rhetoric of culturally led urban development, *Urban Studies* 42 (5–6): 889–911.

Miles, S. and Paddison, R. (2005) Introduction: the rise and rise of culture-led urban regeneration, *Urban Studies* 42 (59–60): 833–9.

Ministry of Sound (2008) *The Story So Far ....* www.ministryofsound.com/sites/fifteen/history (accessed 10 February 2008).

Mintel (2003) *Stag and Hen Holidays, UK.* London: Mintel Group.

Mintel (2004) *High Street Pubs and Bars-UK-April 2004.* London: Mintel Group.

Mintel (2006a) *Dining Out.* London: Mintel Group.

Mintel (2006b) *Eating Out: Ten Year Trends.* London: Mintel Group.

Mintel (2008) *Pub Visiting – UK – September 2008.* London: Mintel Group.

Mitchell, N. (2008) Blood spills in our city of fear, *The Herald Sun*, 26 February. www.news.com.au/heraldsun/story/0,21985,23274713–5000106,00.html (accessed 12 June 2008).

Monclus, F.J. (2003) The Barcelona model: an original formula? From 'reconstruction' to strategic urban projects, *Planning Perspectives* 18: 399–421.

Montemurro, B. (2001) Strippers and screamers: the emergence of social control in a non-institutionalized setting, *Journal of Contemporary Ethnography* 30 (3): 275–304.

Montemurro, B. (2003) Sex symbols: the bachelorette party as a window to change in women's sexual expression, *Sexuality & Culture*, 7 (2): 3–29.

Montemurro, B. (2005) Add men, don't stir: reproducing traditional gender roles in modern wedding showers, *Journal of Contemporary Ethnography* 34 (1): 6–35.

Montgomery, J. (1994) *The Evening Economy of Cities.* The 24-Hour City: Selected Papers from the First National Conference on the Night-time Economy, Manchester.

Montgomery, J. (1995) The story of Temple Bar: creating Dublin's cultural quarter, *Planning Practice and Research* 10 (2): 135–72.

Montgomery, J. (1997) Café Culture and the city: the role of pavement cafés in urban public social life, *Journal of Urban Design* 1 (1): 83–102.

Montgomery, J. (1998) Making a city: urbanity, vitality and urban design, *Journal of Urban Design* 3 (1): 93–116.

Montgomery, J. (2003) Cultural quarters as mechanisms for urban regeneration. Part 1: conceptualising cultural quarters, *Planning Practice and Research* 18 (4): 293–306.

Montgomery, J. (2004a) Born to binge: a rejoinder (writeback), *Town and Country Planning*, 4 December 73 (12): 370.

Montgomery, J. (2004b) Born to Binge? (licensing and the 24 hour city), *Town and Country Planning*, 1 March 73 (3): 82–3.

Montgomery, J. (2007) *The New Wealth of Cities: City Dynamics and the Fifth Wave*. Aldershot: Ashgate.

Moore, C. (2007) *Liquor Amendment (Small Bars and Restaurants) Bill 2007*. Legislative Council. www.parliament.nsw.gov.au/prod/parlment/hansart.nsf/V3Key/LA20070927006 (accessed 14 August 2008).

Moore, C. (2008) New civilised liquor laws, *Clover's eNews* 403. www.clovermoore.com/main/?id=1664 (accessed 17 September 2008).

Moore, R. (2000) Let this city swing for 24 hours a day, *Evening Standard*, 2 August.

Moran, L. and Skeggs, B. (2004) *Sexuality and the Politics of Violence and Safety*. London: Routledge.

Mort, F. (1995) Archaeologies of city life: commercial culture, masculinity, and spatial relations in 1980s London, *Environment and Planning D: Society and Space* 13 (5): 573.

MP Consultancy (2008) *Licensing Law and the Impact of the Public Health Objective: A Review Paper*, Glasgow MP Consultancy. www.alcohol-focus-scotland.org.uk/pdfs/Licensing%2520and%2520Public%2520Health%2520Review.pdf (accessed 21 November 2008).

Munro, G. (2008) Nielsen data shows alcopop tax is working well, *Grogwatch* 24. www.caan.adf.org.au/newsletter.asp?ContentId=t20080804 (accessed 20 November 2008).

Myerscough, J. (1988) *The Economic Importance of the Arts in Britain*. London: Policy Studies Institute.

Nathan, M. and Urwin, C. (2005) *City People: City Centre Living in the UK*. London: Institute for Public Policy Research. www.ippr.org/centreforcities (accessed 13 March 2008).

National UK BIDs Advisory Service (2008) *www.ukbids.org/index.php* (accessed 25 February 2008).

Newburn, T. and Shiner, M. (2001) *Teenage Kicks? Young People and Alcohol: A Review of the Literature*. York: Joseph Rowntree Foundation.

Newton, A., Sarker, S.J., Pahal, G.S., van der Bergh, E. and Young, C. (2007) Impact of the new UK licensing law on emergency hospital attendances: a cohort study, *Emergency Medicine Journal* 24: 532–4. http://emj.bmj.com/cgi/content/full/24/532 (accessed 19 November 2007).

Nicholls, J.Q. (2006) Liberties and licenses: alcohol in liberal thought, *International Journal of Cultural Studies* 9 (2): 131–51.

Nicholson Committee (2003) *A Review of the Liquor Licensing Law in Scotland*. www.scotland.gov.uk/Resource/Doc/47133/0027021.pdf (accessed 21 November 2008).

Nixon, S. and du Gay, P. (2002) Who needs cultural intermediaries?, *Cultural Studies* 16 (4): 495–500.

Norfolk, A. (2007) How 'safe drinking' experts let a bottle or two go to their heads, *The Times*, 20 October.

Nottingham Leisure Business Improvement District (2009) *Go to Nottingham: Live the City Night Life*. www.gotonottingham.co.uk/Introduction-to-Nottingham-Leisure-BID.html (accessed 16 March 2009).

Nottinghamshire Police (n.d.) *The Licensing Act 2003: Advice to Applicants*. www.nottinghamshire.police.uk/uploads/library/18/licensing.pdf (accessed 11 July 2008).

NSW Audit Office (2008) *Working with Hotels and Clubs to Reduce Alcohol-related Crime: Executive Summary*. www.audit.nsw.gov.au/publications/reports/performance/2008/alcohol/execsum.htm (accessed 11 July 2008).

Bibliography

NSW Office of Liquor, Gaming and Racing (2007) *New Liquor Laws*. Office of Liquor, Gaming and Racing.

NSW Office of Liquor Gaming and Racing (2008) *Liquor Regulation: Under the Liquor Act 2007*. Office of Liquor, Gaming and Racing.

Object (2008) *A Growing Tide: Local Authorities Restricted by Inadequate Licensing Laws for Lap Dancing Clubs*. London: Object. www.object.org.uk/downloads/A_Growing_Tide2008.pdf (accessed 29 October 2008).

Oc, T. and Tiesdell, S. (1997) *Safer City Centres: Reviving the Public Realm*. London: Paul Chapman.

ODPM (2004a) *Creating Local Development Frameworks: A Companion Guide to PPS12*. London: The Stationery Office.

ODPM (2004b) *Safer Places: The Planning System and Crime Prevention*. www.communities. gov.uk/publications/planningandbuilding/saferplaces (accessed 27 July 2007).

ODPM (2005a) *Explanatory Memorandum to The Town and Country Planning (General Permitted Development) (Amendment) (England) Order 2005*. London: The Stationery Office.

ODPM (2005b) *Planning Policy Statement 6: Planning for Town Centres*. London: The Stationery Office.

ODPM (2005c) *How to Manage Town Centres*. London: ODPM.

Ofcom (2004) *Alcohol and Advertising: Consultation Document*. London: Office of Communications.

Office of National Statistics (2003) *Households: One-person Households up to 30%*. www.stat-istics.gov.uk/cci/nugget.asp?id=350 (accessed 14 August 2008).

Office of National Statistics (2004) *Drinking: Adults' Behaviour and Knowledge in 2004*. London: Office of National Statistics.

Office of National Statistics (2007a) *Households and Families: Highlights*. www.statistics.gov.uk/ CCI/nugget.asp?ID=1748&Pos=1&ColRank=2&Rank=224 (accessed 21 October 2008).

Office of National Statistics (2007b) *Statistics on Alcohol: England 2007*. www.ic.nhs.uk/ cmsincludes/_process_document.asp?sPublicationID=1180968573656&sDocID=3493(accessed 24 November 2008).

Office of National Statistics (2008a) *Alcohol Deaths: Rates in the UK Continue to Rise*. www. statistics.gov.uk/CCI/nugget.asp?ID=1091&Pos=1&ColRank=1&Rank=192, (accessed) 25 July 2008).

Office of National Statistics (2008b) *Households and Families: Cohabiting Increases*. www.stat-istics.gov.uk/cci/nugget.asp?id=1925 (accessed 8 June 2008).

Oldenburg, G. (1999) *The Great Good Place: Cafés, Coffee Shops, Bookstores, Hair Salons and Other Hangouts at the Heart of a Community*. New York: Marlowe and Company.

Olsen, D.J. (1986) *The City as a Work of Art: London, Paris, Vienna*. New Haven: Yale University Press.

O'Malley, P. and Valverde, M. (2004) Pleasure, freedom and drugs: the uses of 'pleasure' in liberal governance of drug and alcohol consumption, *Sociology* 38 (1): 25–42.

Oosterman, J. (1994) *Café Culture, Urban Space and the Public Realm*. The 24-Hour City: Selected Papers from the First National Conference on the Night-time Economy, Manchester.

Open All Hours? Campaign (2002) *Open All Hours? A Report on Licensing Deregulation*. London: The Civic Trust and the Institute of Alcohol Studies.

Pain, R. (2001) Gender, race, age and fear in the city, *Urban Studies* 38 (5–6): 899–913.

Painter, K.A. and Farrington, D. (2001) The financial benefits of improved street lighting, based on crime reduction, *Lighting Research & Technology* 33 (1): 3–12.

Palmer, B. (2000) *Cultures of Darkness: Night Travels in the History of Transgression: From Medieval to Modern*. New York: Monthly Review Press.

Parker, H. and Williams, L. (2003) Intoxicated weekends: young adults work hard–play hard life-styles, public health and public disorder, *Drugs: Education, Prevention & Policy* 10 (4): 345–7.

Parkes, D.N. and Thrift, N.J. (1980) *Times, Spaces, and Places: A Chronogeographic Perspective*. Chichester: John Wiley & Sons.

Pincock, S. (2003) Binge drinking on rise in UK and elsewhere, *Lancet* 362: 1126–7.

Pini, M. (2001) *Club Cultures and Female Subjectivity: The Move from Home to House*. Basingstoke: Palgrave.

Planning Transport and Highways (2001) *Devonshire Quarter Action Plan*. Sheffield: Sheffield City Council.

Plant, E.J. and Plant, M. (2005) A 'leap in the dark?' Lessons for the United Kingdom from past extensions of bar opening hours, *International Journal of Drug Policy* 16 (6): 363–8.

Plant, M. and Plant, M. (2006) *Binge Britain: Alcohol and the National Response*. Oxford: Oxford University Press.

Portman Group (2008) Factsheet. www.portman-group.org.uk/?pid=27&level=3 (accessed 27 November 2008).

Prime Minister's Strategy Unit (2003) *Interim Analytic Report*. Cabinet Office.

Quinn, P.E. (1996) *Temple Bar: The Power of an Idea*. Dublin: Temple Bar Properties Ltd.

Raban, J. (1998) *Soft City*. London: Harvill.

Ramsay, M. (1990) *Lagerland Lost? An Experiment in Keeping Drinkers off the Streets in Central Coventry and Elsewhere*. London: Home Office, Crime Prevention Unit.

Redden, G. (2008) The great British binge drinking debate, *Soundings: A Journal of Politics and Culture* 39: 117–27.

Reynolds, C. (1999) Dublin's Cinderella neighbourhood, *Florida Times*, 5 September.

Reynolds, E.A. (1998) *Before the Bobbies: The Night Watch and Police Reform in Metropolitan London 1720–1830*. Stanford: Stanford University Press.

Richardson, A. and Budd, T. (2003) *Alcohol, Crime and Disorder: A Study of Young Adults*. Home Office Research Study No. 263. London: Home Office.

Richardson, A., Budd, T., Engineer, R., Phillips, A., Thompson, J. and Nicholls, J. (2003) *Drinking, Crime and Disorder*. London: Home Office.

Robbins, T. (2005) Focus: do you want this 24 hours a day?, *Sunday Times*, 16 January. www.timesonline.co.uk/tol/news/uk/article413019.ece?token=null&offset=12&page=2 (accessed 29 October 2008).

Roberts, M. (1998) Urban design and regeneration. In Greed, C. and Roberts, M. (eds) *Introducing Urban Design: Interventions and Responses*. Harlow: Longman, pp. 149–78.

Roberts, M. (2004) *Good Practice in Managing the Evening and Late-Night Economy: A Literature Review from an Environmental Perspective*. London: ODPM.

Roberts, M. and Gornostaeva, G. (2007) The night-time economy and sustainable city centres: dilemmas for local government, *International Journal of Sustainable Development and Planning* 2 (2): 1–19.

Roberts, M. and Turner, C. (2005) Conflicts of liveability in the 24-hour city: learning from 48 hours in London's Soho, *Journal of Urban Design* 10 (2): 171–93.

Roberts, M., Turner, C., Greenfield, S. and Osborn, G. (2006) A continental ambience? Lessons in managing alcohol related evening and night-time entertainment from four European capitals, *Urban Studies* 43 (7): 1105–25.

Rood, D. (2007) Tough laws aim at CBD violence, *The Age*, 31 August 31. www.theage.com.au/news/national/tough-laws-aim-at-cbd-violence/2007/08/30/1188067278015.html (accessed 31 August 2007).

Room, R. (2004) Disabling the public interest: alcohol strategies and policies for England, *Addiction* 99 (9): 1083–9.

Rowan, D. (1998) The boom in one-person households, *Guardian*, 20 January.

Rudd, K. (2008) *National Binge Drinking Strategy*. www.labor.com.au/media/0308/mspm100.php.

Schivelbusch, W. (1988) *Disenchanted Night: The Industrialisation of Light in the Nineteenth Century*. Oxford: Berg.

Schlör, J. (1998) *Nights in the Big City: Paris, Berlin, London, 1840–1930*. London: Reaktion Books Ltd.

Bibliography

Scottish Government (2008) *Changing Scotland's Relationship with Alcohol: A Discussion Paper On Our Strategic Approach*. Edinburgh: Scottish Government.

Secretary of State for Culture (2002) *Guidance Issued under Section 177 of the Licensing Act 2003 v.2.0*. London: Department for Culture, Media and Sport.

Secretary of State for the Department for Culture, Media and Sport (2004) *Draft Guidance Issued Under Section 182 of the Licensing Act 2003. Tabled before Parliament on 23 March 2004*. London: Department for Culture, Media and Sport.

Sennett, R. (1996) *Flesh and Stone: The Body and the City in Western Civilization*. New York: W.W. Norton.

Sennett, R. (1999) *The Corrosion of Character: The Personal Consequences of Work in the New Capitalism*. London: W.W. Norton.

Sennett, R. (2007) The open city. In Burdett, R. and Sudjic, D. (eds) *The Endless City*. London: Phaidon, pp. 290–7.

Shaw, S. and Karmowska, J. (2004) Ethnoscapes as spectacle: re-imaging muticultural districts as new destinations for leisure and tourism consumption, *Urban Studies* 41 (10): 1983–2000.

Shorthose, J. (2004) Nottingham's de facto cultural quarter: the lace market, independents and a convivial ecology. In Bell, D. and Jayne, D. (eds) *City of Quarters: Urban Villages and the Contemporary City*. London: Ashgate.

Simms, A., Kjell, P. and Potts, R. (2005) *Clone Town Britain*. London: New Economics Foundation.

Sivarajasingam, V., Shepherd, J.P. and Matthews, K. (2003) Effect of closed circuit television on assault injury and violence detection, *Injury Prevention* 9 (4): 312–16.

Skeggs, B. (2005) The making of class and gender through visualizing moral subject formation, *Sociology* 39 (5): 965–82.

Slack, J. (2005) Jowell slapped down by police over drinks laws, *Daily Mail*, 16 November, 6–7.

Smith, A. (2005) Conceptualizing city image change: the re-imaging of Barcelona, *Tourism Geographies* 7 (4): 398–423.

Smith, A. (2007a) Moore fails to convince liberals on liquor laws, *Sydney Morning Herald*, 4 September. www.smh.com.au/news/national/moore-fails-to-convince-liberals-on-liquor-laws/2007/08/23/1187462441729.html (accessed August 24 2008).

Smith, G.D. (2008) The night-time economy: exploring tensions between agents of control. In Atkinson, R. and Helms, G. (eds) *Securing an Urban Renaissance: Crime, Community and British Urban Policy*. Bristol: Policy Press, pp. 183–202.

Smith, M.A. (1983) Social usages of the public drinking house: changing aspects of class and leisure, *British Journal of Sociology* 34 (3): 367–85.

Smith, N. (1996) *The New Urban Frontier: Gentrification and the Revanchist City*. London: Routledge.

Smith, R. (2007b) A row plucked out of the air, *Guardian*, 22 October. http://commentisfree.guardian.co.uk/richard_smith/2007/10/a_row_plucked_out_of_the_air.html (accessed 23 October 2007).

Social Issues Research Centre (2002) *Counting the Cost: The Measurement and Recording of Alcohol-related Violence and Disorder*. London: The Portman Institute. www.sirc.org/publik/alcohol_related_violence4.shtml (accessed 13 April 2008).

Somers, D. (1971) The leisure revolution: recreation in the American City, 1820–1920, *Journal of Popular Culture* 5 (Summer): 125–47.

Sparks, R., Girling, E. and Loader, I. (2001) Fear and everyday urban lives, *Urban Studies* 38 (5–6): 885–98.

Standing Committee on Community Affairs (2008) *Ready-to-drink Alcohol Beverages*, Canberra: The Committee.

Statham, J. and Mooney, A. (2002) *Around the Clock: Childcare Services at Atypical Times*. York: Joseph Rowntree Foundation Policy Press.

Stevens, Q. (2007) *The Ludic City: Exploring the Potential of Urban Spaces*. Abingdon: Routledge.

Summers, A. (2008) Binge drinking something to wine about, *Sydney Morning Herald.* www.smh. com.au/news/opinion/binge-drinking-something-to-wine-about/2008/06/20/1213770920995. html?page=fullpage#contentSwap2 (accessed 20 October 2008).

Talbot, D. (2004) Regulation and racial differentiation in the construction of night-time economies: a London case study, *Urban Studies* 41 (4): 887–902.

Talbot, D. (2006) The Licensing Act 2003 and the problematization of the night-time economy: planning, licensing and subcultural closure in the UK, *International Journal of Urban and Regional Research* 30 (1): 159–71.

Talbot, D. (2007) *Regulating the Night: Race, Culture and Exclusion in the Making of the Night-time Economy.* Aldershot: Ashgate.

Talbot, D. and Bose, M. (2007) Racism, criminalization and the development of night-time economies: two case studies in London and Manchester, *Ethnic and Racial Studies* 30 (1): 95–118.

Tallon, A.R., Bromley, R., Reynolds, B. and Thomas, C.J. (2006) Developing leisure and cultural attractions in the regional city centre: a policy perspective, *Environment and Planning C: Government and Policy* 24: 351–70.

Taylor, B. and Hughes, D. (2005) Police chiefs: we're against 24-hour pubs, *Daily Mail*, 13 January, 1–2.

Thomas, C.J. and Bromley, R.D.F. (2000) City-centre revitalisation: problems of fragmentation and fear in the night-time city, *Urban Studies* 37 (8): 1403.

Thornley, A. (1991) *Urban Planning under Thatcherism: The Challenge of the Market.* London: Routledge.

Thornton, S. (1995) *Club Cultures: Music, Media and Subcultural Capital.* Cambridge: Polity Press.

Tiesdell, S. and Slater, A.-M. (2006) Calling time: managing activities in space and time in the evening/night-time economy, *Planning Theory and Practice* 7 (2): 137–57.

Toffler, A. (1971) *Future Shock.* London: Pan Books.

Törrönen, J. (2001) Between public good and the freedom of the consumer: negotiating the space, orientation and position of us in the reception of alcohol policy editorials, *Media, Culture and Society* 23 (2): 171–93.

Törrönen, J. and Karlsson, T. (2005) Moral regulation of public space and drinking in the media and legislation in Finland, *Contemporary Drug Problems* 32: 93–126.

Town Centres Ltd (2001) *West End Entertainment Impact Study.* London: Westminster City Council.

Treadwell, J. (2008) Binge drinking and young people: a potent combination?, *Birmingham Post.* http://blogs.birminghampost.net/news/2008/03/binge-drinking-and-young-peopl.html (accessed 4 March 2008).

Tye, D. and Powers, A.M. (1998) Gender, resistance and play: bachelorette parties in Atlantic Canada, *Women's Studies International Forum* 21 (5): 551–61.

UK Parliament (n.d.) *Explanatory Memorandum to the Local Authorities (Alcohol Disorder Zones) Regulations.* www.opsi.gov.uk/si/si2008/draft/em/ukdsiem_9780110808093_en.pdf (accessed 27 November 2008).

Unsworth, R. (2007) *City Living in Leeds: 2007.* Leeds: University of Leeds.

Urban Practitioners (2004) *Hackney Night-Time Economy Evidence Based Study: Shoreditch.* London: London Borough of Hackney.

Urban Splash (2008) *Our Story.* www.urbansplash.co.uk/Default.aspx?id=152&area=13 (accessed 28 October 2008).

Urban Task Force (1999) *Towards an Urban Renaissance.* London: E & FN Spon.

Urbed, CASA and Lovatt, A. (2002) *Late-Night London: Planning and Managing the Late Night Economy.* London: University College London, Greater London Authority, London Development Agency and Transport for London.

Valanecia-Martin, J.L., Galan, Inaki and Rodriguez-Artalejo, F. (2007) Binge drinking in Madrid, Spain, *Alcoholism: Clinical and Experimental Research* 31 (10): 1723–30.

Bibliography

Valentine, G.S., Holloway, S., Knell, C. and Jayne, M. (2008) Drinking places: young people and cultures of alcohol consumption in rural environments, *Journal of Rural Studies* 24 (1): 28–40.

Van den Berg, L., Pol, P.M.J., Mingardo, G. and Speller, C.J.M. (2006) *The Safe City: Safety and Urban Development in European Cities*. Aldershot: Ashgate.

Watson, S. and Gibbs, K. (1995) *Postmodern Cities and Spaces*. Oxford: Blackwell.

Webster, R. (2008) *Safer Nightlife: Best Practice for those Concerned about Drug Use and the Night-time Economy*. http://drugs.homeoffice.gov.uk/news-events/latest-news/safernightlife (accessed 25 July 2008).

Webster, R., Goodman, M. and Whally, G. (2002) *Safer Clubbing Guidance for Licensing Authorities, Clubmanagers and Promoters*. London: The Home Office and London Drugs Forum in partnership with Release. www.crimereduction.homeoffice.gov.uk/drugsalcohol/drugsalcohol49.htm (accessed 25 July 2008).

Welsh, B.C. and Farrington, D.C. (2002) *Crime Prevention Effects of Closed Circuit Television: A Systematic Review*. London: The Home Office.

Welsh, B.C. and Farrington, D. (2006) Surveillance for crime prevention in public space: results and policy choices in Britain and America, *Criminology and Public Policy* 3 (3): 497–526.

Westminster City Council (1997) *Analysis of Planning Policies (January 1992–December 1996)*. www.westminster.gov.uk/udp/adopted/monrep (accessed 21 July 2008).

Westminster City Council (2001) *Second Deposit Unitary Development Plan*. London: Department of Planning.

Westminster City Council (2003) *Leicester Square: Al Fresco*, information sheet.

White, M. and Hetherington, P. (2005) Ministers at war over pub closing time, *Guardian*, 15 January, 1.

Whyte, W.H. (1980) *The Social Life of Small Urban Spaces*. Washington: The Conservation Foundation.

Williams, R. (2008) Darkness, deterritorialization, and social control, *Space and Culture* 11 (4): 514–32.

Williamson, E. (2006) Only here for the beer, *Travel Weekly: The Choice of Travel Professionals*, 16 June 1825.

Wilson, B. (2006) *Fight, Flight or Chill: Subcultures, Youth, and Rave into the Twenty-First Century*. Montreal: McGill-Queen's University Press.

Wilson, T.M. (ed.) (2005) *Drinking Cultures: Alcohol and Identity*. Oxford: Berg.

Winlow, S. and Hall, S. (2006) *Violent Night: Urban Leisure and Contemporary Culture*. Oxford: Berg.

Wittel, A. (2001) Toward a network sociality, *Theory, Culture & Society* 18 (6): 51–76.

Woollett, V. (2008) *Investigating the Role of the Planning System in Supporting Independents in the Night-time Economy: A Comparison of Two Night-time Economies; The Angel, Islington and Shoreditch, Hackney*. London: Urban Development and Regeneration, University of Westminster.

Working Solutions (1998) *Keeping the Peace: A Guide to the Prevention of Alcohol Related Disorder*. London: The Portman Group.

Worpole, K. (1992) *Towns for People*. Milton Keynes: Open University Press.

Worpole, K. and Greenhalgh, L. (1999) *The Richness of Cities: Final Report*. London: Comedia and Demos.

Worpole, K. and Knox, K. (2007) *The Social Value of Public Spaces*. York: Joseph Rowntree Foundation. www.jrf.org.uk/bookshop/eBooks/2050-public-space-community.pdf (accessed 22 July 2008).

Wostear, S. (2007) It's our poison … so leave us alone. Sun man's defence of 'danger' wine tipplers, *Sun*. www.thesun.co.uk/sol/homepage/news/article351841.ece (accessed 17 October 2007).

Zukin, S. (1995) *The Cultures of Cities*. Oxford: Blackwell.

Index

24 hour city/society 21, 25–6, 34–9,
    138
24 hour drinking 2, 114

advertising 103–4
alcohol: and actual consumption
    88–91, 97; health 87–8, 92–3;
    identity 186–7; alternatives to
    drinking 170, 196; *see also* leisure
    centres; eating out; shopping;
    festivals
alcohol disorder zones 177–8
alcohol free zones 167
*Alcohol Harm Reduction Strategy,
    England* (AHRSE) 83, 107, 159, 173
alcopops 78, 83, 96, 105, 151
anonymity 18
anti-social behaviour 2, 16, 63–72,
    115, 150, 166–7
Association of Chief Police Officers
    (ACPO) 1
Australia 105, 151–2; *see also* New
    South Wales; Melbourne,
    Australia

Barcelona 9, 40; small bars 151
Barcelona 'model' 44, 143–51
bars: pubs and bars 19, 187
Berlin, Germany 181
Best Bar None 164
Better Regulation Taskforce 111
binge Britain 16, 82, 95–7
binge drinking 82–108, 218–19;
    as anti-youth 84, 100; as gendered
    100; problems with definition
    84–8; reasons for
    102–3, 188
bottellon 99
brewers 45; *see also* pubcos
Business Improvement Districts
    176–7

café culture 38, 138, 221, 205
CAMRA (Campaign for Real Ale) 45,
    193–4
capacity 46, 52, 148, 172, 193
CCTV 165–6
chain-bars 50–2, 192–3, 195; *see
    also* independent venues

change of use 49, 140
Chatterton, P. and Hollands, R. 4, 7,
    32, 146, 173
children 8, 200–1
cinema 8, 197–8
city centre living 42–4, 64–9
Civic Trust 160, 170
Comedia 35–8
concentration of venues 60, 62, 81,
    91, 116, 117, 129, 177, 213–14
Copenhagen, Denmark 8, 139–43,
    181
creative industries 3
crime 16, 91–2, 116–17, 191; and
    Licensing Act 2003 126–9
cultural quarters and culture led
    regeneration 39–41, 44; Temple
    Bar 74–7; Helsinki 79–80
cumulative impact 121; in Barcelona
    151
Cumulative Impact Areas 131, 182–3

deregulation 5, 9
dispersal 118, 120–1, 214
door staff 165
drinking circuit 5
drinking streets 61
drugs 102–4, 172
drunkenness 2, 87, 90, 95, 109,
    186

eating 25; and eating out 45,
    198–299
economic impact 57
Edinburgh 9, 64, 70–2, 73, 120, 172,
    176
education; expansion of 42; *see also*
    students
enforcement 149, 164
environmental improvements 36
Europe and alcohol consumption
    97–9; as idealized 56, 136–58,
    220–1
European ideal 137–9
everyday, the 12, 21, 187–8, 212
expansion of night-time economy
    58–9, 65–8, 78; in Copenhagen
    143
extended hours 125

families 137, 196–7, 200–2
fast-food 129
fear/pleasure binary 12, 13
fear 15–17, 69, 191
festivals 79–80, 201–2
Florida, R. 40–1, 142

gender relations 22, 28, 31, 42,
    102
gentrification 41, 44–5, 210–11; in
    Copenhagen 141
globalization 25–6
Guidance 115, 121, 124, 131

Hackney, East London 43
Hadfield, P. 2–3, 5, 110, 111, 121,
    161
happy hours 37–8, 114, 133
harmful drinking 88
hazardous drinking 88
health 92–3, 134
Helsinki, Finland 9, 77–80
hen parties 72–3, 76, 80, 145
heterotopia 56, 80
higher education 30
Hobbs, D. 1, 6, 66, 91, 101, 114,
    161–2
Holloway, S. 94

independent venues 111, 193–6,
    204–5, 220
Individualization 27; *see also*
    responsibilization

Jacobs, J. 34–5
Jayne, M. 5, 11

Kreitzman, L. 22

lap-dancing 131–3
Latham, A. 11, 44, 186, 192, 222
Leeds 60; and urban renaissance
    42–3
leisure 17–20, 190
leisure centres 196–8
lesbian and gay culture 17, 186
licensing, variations to hours 38
Licensing Act 1964 110
Licensing Act 1988 45, 110–11

Index

Licensing Act 2003 2, 8, 113, 119:
and impact on crime 126–9;
evidence on which it was based
116–21; extended hours 125;
democracy 123; local councils
123–4, 130–1; media reception
114–15; see also Time for Reform
licensing conditions 124
licensing forums 174–6
licensing hearings 130
licensing objectives 113, 215
licensing policies 131
Licensing Standards Officer 133
lighting 13–15, 170
liminality 17, 55–7, 162
litter 66, 69, 169
Liverpool and the urban renaissance
42–3
local authorities 113, 122, 169–70
London 19
Lumina plc 47

magistrates 111, 124
Manchester 42–3, 46–7, 60, 62,
163–5; Hacienda club 46
Melbin, M. 15, 16–18, 21–2
Melbourne, Australia 16, 152–4, 155;
small bars 152–3, 193
Metropolitan Police 16, 119
migration (of land uses) 121–2
mixed use 42, 44, 122, 144, 147,
181–2, 221
Montgomery, J. 2, 5, 41, 73, 74
moral panic 96, 100

need, concept of 49, 110
Netherlands 118
New South Wales 152, 193; Liquor
Act 2007 83, 152–6
Newcastle-upon-Tyne 60
Newcastle-under-Lyme 65
nightlife 1, 18, 20, 40
night-time economy 1, 2, 10, 11,
23–4, 142, 161; and evening 41,
116
no-go areas 1, 160
noise 64–5, 67, 143, 147, 169
Norwich 5, 127–9, 173
Nottingham 60, 177

off-licenses; see also supermarkets;
public/private drinking; tension
with on-trade 83

over-provision zone 71; see also
Cumulative Impact Zone;
saturation zone

partnerships 173–8; see also
Business Improvement Districts
pavement cafes 170
pedestrian density 129
pedestrianization 36–7, 139, 213
Planning and Compulsory Purchase
Act 181
planning conditions 180
pleasure 18, 189, 207
policing 118, 163–4, 168
post-industrialization 33–4
pre-loading 5, 106
pricing 104–7
pubcos 45, 47, 50–2
public/private drinking 78, 83, 94–5
public space 139, 144, 149–50, 162,
196
public urination 169–70
pubs 4, 20, 30, 36, 61–2, 193–5, 205,
219–20
pubwatch 165, 177

race and ethnicity 6
regeneration 46–7, 141–2, 199
residents 56, 64–6, 69, 71, 76, 79, 81
responsibilization 84, 101–3, 169
restaurants 45, 51–2, 198–200; see
also Melbourne
Rudd, Kevin 83, 151

Safe. Sensible. Social 87; see also
Alcohol Harm Reduction Strategy
saturation 121
Schivelbusch, W. 13–15, 18
Schlör, J.11, 16, 17
Scotland 106; Licensing Act 133–4;
see also Edinburgh
sex 55
shopping 170, 202–3
Shoreditch, London 195
single households 28–30, 79
small bars see Melbourne and New
South Wales and Barcelona
smoking 195–6, 200
sociality 17, 19, 27–8, 55; network
sociality 24–5, 41
Soho, London 7, 56, 66–70
Spain 98–9, 149
special policy area 121, 131; see also

Cumulative Impact Zone;
Cumulative Impact Areas
stag parties see hen parties
street café 38
students 28, 30, 42, 70–1, 79
supermarkets 83, 94, 105–6; see
also public/private drinking
supply chains 38
Supply of Beer Orders Act 45, 48
Swansea, England 61
Sydney, Australia see New South
Wales

taxis 171
Temple Bar, Dublin 8, 73–7, 181
temporal divisions 60–1
terminal hours 37, 45, 110–11,
117–20, 125–6, 157; see also
Copenhagen; Barcelona;
Melbourne
theatres 8, 18, 20
Time for Reform 112–13
tourism 111, 137, 142, 145
town centre management 7
town planning 4, 178–82, 188,
203–4, 211–12, 214–15
transport 171–2, 213

under-age drinking 148, 168
United States 56; definition of binge
drinking 86
units (of alcohol) 85–7
urban renaissance 6, 41–5, 64, 66,
144, 192
Urban Task Force 42, 44–5
urbanity 4, 17
use class orders 122, 182

Valentine, G. 17
violence 191; and alcohol 62–4; and
closing times 37

West End stress area 67
Westminster City Council 67, 121,
162, 169–70, 180–1
women 6, 30–2, 91, 100
work 23–5, 57–8; changing patterns
of work 20

youth 17, 84, 89,-90, 101, 103–4; and
dominance of nightlife 60–2; and
urban/rural 17